METHODS IN MOLECULAR BIOLOGY™

Series Editor
John M. Walker
School of Life Sciences
University of Hertfordshire
Hatfield, Hertfordshire, AL10 9AB, UK

For further volumes:
http://www.springer.com/series/7651

Chemokines

Methods and Protocols

Edited by

Astrid E. Cardona

Department of Biology, South Texas Center for Emerging Infectious Diseases, The University of Texas at San Antonio, San Antonio, TX, USA

Eroboghene E. Ubogu

Department of Neurology, Baylor College of Medicine, Houston, TX, USA

Humana Press

Editors
Astrid E. Cardona
Department of Biology
South Texas Center for Emerging Infectious Diseases
The University of Texas at San Antonio
San Antonio, TX, USA

Eroboghene E. Ubogu
Department of Neurology
Baylor College of Medicine
Houston, TX, USA

ISSN 1064-3745 ISSN 1940-6029 (electronic)
ISBN 978-1-4939-6005-7 ISBN 978-1-62703-426-5 (eBook)
DOI 10.1007/978-1-62703-426-5
Springer New York Heidelberg Dordrecht London

Printed on acid-free paper

Humana Press is a brand of Springer
Springer is part of Springer Science+Business Media (www.springer.com)

Preface

Chemokines constitute a large family of structurally similar cytokines that contain a signature of conserved cysteine residues joined by disulfide bridges. Binding of chemokines to specific G-protein-coupled receptors followed by downstream signaling defines their biological function. Initially, chemoattraction was the key function linked to chemokines/chemokine receptors. However, in recent years, it has become clear that chemokine ligand–receptor interactions can also modulate cellular activation, survival, and proliferation, among other functions in homeostatic and diseased states. Importantly, major advances in our understanding of chemokine biology have led to chemokine receptors becoming specific therapeutic targets with great potential. This book provides practical information regarding experimental models and state-of-the-art protocols used to delineate chemokine/chemokine receptor function and their applications in health and disease. The reader is exposed to step-by-step protocols of functional and descriptive chemokine ligand/receptor assays that will be of practical significance to graduate students, postdoctoral fellows, trainees, and researchers in academia and industry.

San Antonio, TX, USA *Astrid E. Cardona*
Houston, TX, USA *Eroboghene E. Ubogu*

Contents

Contributors

RUBEN ABAGYAN • *Skaggs School of Pharmacy and Pharmaceutical Sciences, University of California San Diego, La Jolla, CA, USA*

KEN ARAI • *Neuroprotection Research Laboratory, Massachusetts General Hospital and Harvard Medical School, Boston, MA, USA; Department of Neurology and Radiology, Clinical Proteomics Research Center, Massachusetts General Hospital and Harvard Medical School, Boston, MA, USA*

ANNE-MARIE BAIRD • *Department of Clinical Medicine, Trinity College, Dublin, Ireland; Thoracic Oncology Research Group, Institute of Molecular Medicine, Trinity Centre for Health Science, St James's Hospital, Dublin, Ireland*

FANNY BELTRAN • *Geneva Research Centre, Division of Lead Discovery Technologies, Merck Serono, Geneva, Switzerland EMD Serono Research Institute, Billerica, MA, USA*

LUIGI BUONAGURO • *Molecular Biology and Viral Oncology, Department of Experimental Oncology, Istituto Nazionale Tumori "Fond G. Pascale", Naples, Italy*

JING CAO • *Neuroprotection Research Laboratory, Massachussetts General Hospital and Harvard Medical School, Boston, MA, USA; Department of Neurology and Radiology, Clinical Proteomics Research Center, Massachusetts General Hospital and Harvard Medical School, Boston, MA, USA*

SUSANNA CARBONI • *Geneva Research Centre, Division of Autoimmune and Inflammatory Diseases, Merck Serono, Geneva, Switzerland*

ASTRID E. CARDONA • *Department of Biology, South Texas Center for Emerging Infectious Diseases, The University of Texas at San Antonio, San Antonio, TX, USA*

SANDRA M. CARDONA • *Department of Biology, South Texas Center for Emerging Infectious Diseases, The University of Texas at San Antonio, San Antonio, TX, USA*

MONICA J. CARSON • *Division of Biomedical Sciences, Center for Glial–Neuronal Interactions, University of California Riverside, Riverside, CA, USA*

GUSTAVO FENALTI • *The Scripps Research Institute, La Jolla, CA, USA*

LAURA B. FORD • *Institute of Infection, Immunity and Inflammation, College of Medical, Veterinary and Life Sciences, University of Glasgow, Glasgow, UK*

ALEXANDRE GARIN • *Geneva Research Centre, Division of Autoimmune and Inflammatory Diseases, Merck Serono, Geneva, Switzerland*

JENNY A. GARCIA • *Department of Biology, South Texas Center for Emerging Infectious Diseases, The University of Texas at San Antonio, San Antonio, TX, USA*

MARILÈNE GAUDET • *Geneva Research Centre, Division of Autoimmune and Inflammatory Diseases, Merck Serono, Geneva, Switzerland*

C. TAYLOR GILLILAND • *Skaggs School of Pharmacy and Pharmaceutical Sciences, University of California San Diego, La Jolla, CA, USA*

STEVEN G. GRAY • *Department of Clinical Medicine, Trinity College, Dublin, Ireland; Thoracic Oncology Research Group, Institute of Molecular Medicine, Trinity Centre for Health Science and HOPE Directorate, St James's Hospital, Dublin, Ireland*

SHUZHEN GUO • *Neuroprotection Research Laboratory, Massachusetts General Hospital and Harvard Medical School, Boston, MA, USA; Department of Neurology and Radiology, Clinical Proteomics Research Center, Massachusetts General Hospital and Harvard Medical School, Boston, MA, USA*

REBECCA HAFFNER • *Department of Immunology, The Weizmann Institute of Science, Rehovot, Israel*

DAMON HAMEL • *Skaggs School of Pharmacy and Pharmaceutical Sciences, University of California San Diego, La Jolla, CA, USA; Janssen Research and Development, San Diego, CA, USA*

TRACY M. HANDEL • *Skaggs School of Pharmacy and Pharmaceutical Sciences, University of California San Diego, La Jolla, CA, USA*

CHRIS A.H. HANSELL • *Institute of Infection, Immunity and Inflammation, College of Medical, Veterinary and Life Sciences, University of Glasgow, Glasgow, UK*

AURELIE HERMANT • *Geneva Research Centre, Division of Autoimmune and Inflammatory Diseases, Merck Serono, Geneva, Switzerland*

HONG JI • *Geneva Research Centre, Division of Autoimmune and Inflammatory Diseases, Merck Serono, Geneva, Switzerland; Novo Nordisk, Màlev, Denmark*

ZOË JOHNSON • *Geneva Research Centre, Division of Autoimmune and Inflammatory Diseases, Merck Serono, Geneva, Switzerland*

STEFFEN JUNG • *Department of Immunology, The Weizmann Institute of Science, Rehovot, Israel*

WILLIAM J. KARPUS • *Department of Pathology, Robert H. Lurie Comprehensive Cancer Center, Interdepartmental Immunobiology Center, Center for Genetic Medicine, Feinberg School of Medicine, Northwestern University, Chicago, IL, USA*

KI-WOOK KIM • *Department of Immunology, The Weizmann Institute of Science, Rehovot, Israel*

EVA KOCHHAN • *Max Planck Institute for Molecular Biomedicine, Münster, Germany*

SONJA KROHN • *Geneva Research Centre, Division of Autoimmune and Inflammatory Diseases, Merck Serono, Geneva, Switzerland*

IRINA KUFAREVA • *Skaggs School of Pharmacy and Pharmaceutical Sciences, University of California San Diego, La Jolla, CA, USA*

OLIVIER LEGER • *Geneva Research Centre, Division of Protein Engineering and Antibody Technologies, Merck Serono, Geneva, Switzerland*

MEIZHANG LI • *Department of Stem Cell Biology and Regenerative Medicine, Lerner Research Institute, Cleveland Clinic Foundation, Cleveland, OH, USA; School of Life Sciences, Yunnan University, China*

ENG H. LO • *Neuroprotection Research Laboratory, Massachusetts General Hospital and Harvard Medical School, Boston, MA, USA; Department of Neurology and Radiology, Clinical Proteomics Research Center, Massachussetts General Hospital and Harvard Medical School, Boston, MA, USA*

OLIMPIA MEUCCI • *Departments of Pharmacology and Physiology and Microbiology and Immunology, Drexel University College of Medicine, Philadelphia, PA, USA*

ALEXANDRA MICHEL • *Geneva Research Centre, Division of Protein Engineering and Antibody Technologies, Merck Serono, Geneva, Switzerland*

MARC MISSOTTEN • *Geneva Research Centre, Division of Lead Discovery Technologies, Merck Serono, Geneva, Switzerland*

ROBERT J.B. NIBBS • *Institute of Infection, Immunity and Inflammation, College of Medical, Veterinary and Life Sciences, University of Glasgow, Glasgow, UK*

MINGMING NING • *Neuroprotection Research Laboratory, Massachussetts General Hospital and Harvard Medical School, Boston, MA, USA; Department of Neurology and Radiology, Clinical Proteomics Research Center, Massachusetts General Hospital and Harvard Medical School, Boston, MA, USA*

KENNETH J. O'BYRNE • *Thoracic Oncology Research Group, Institute of Molecular Medicine, Trinity Centre for Health Science and HOPE Directorate, St James's Hospital, Dublin, Ireland*

JONATHAN PITCHER • *Departments of Pharmacology and Physiology and Microbiology and Immunology, Drexel University College of Medicine, Philadelphia, PA, USA*

CHRISTINE POWER • *Geneva Research Centre, Division of Protein and Antibody Technologies, Merck Serono, Geneva, Switzerland*

AMANDA PROUDFOOT • *Geneva Research Centre, Division of Autoimmune and Inflammatory Diseases, Merck Serono, Geneva, Switzerland*

YANN RATINAUD • *Geneva Research Centre, Division of Lead Discovery Technologies, Merck Serono, Geneva, Switzerland*

SAORI SHIMIZU • *Department of Pharmacology and Physiology, Drexel University College of Medicine, Philadelphia, PA, USA*

ARNDT F. SIEKMANN • *Max Planck Institute for Molecular Biomedicine, Münster, Germany*

MICHAEL SIXT • *Institute of Science and Technology Austria, Klosterneuburg, Austria*

BRYAN STEPHENS • *Skaggs School of Pharmacy and Pharmaceutical Sciences, University of California San Diego, La Jolla, CA, USA*

RAYMOND C. STEVENS • *The Scripps Research Institute, La Jolla, CA, USA*

MARIA TAGLIAMONTE • *Molecular Biology and Viral Oncology, Department of Experimental Oncology, Istituto Nazionale Tumori "Fond G. Pascale", Naples, Italy*

EROBOGHENE E. UBOGU • *Neuromuscular Immunopathology Research Laboratory, Department of Neurology, Baylor College of Medicine, Houston, TX, USA*

MARIA LUISA VISCIANO • *Molecular Biology and Viral Oncology, Department of Experimental Oncology, Istituto Nazionale Tumori "Fond G. Pascale", Naples, Italy*

MICHELE WEBER • *Institute of Science and Technology Austria, Klosterneuburg, Austria*

EMMA H. WILSON • *Division of Biomedical Sciences, Center for Glial-Neuronal Interactions, University of California Riverside, Riverside, CA, USA*

BEILI WU • *The Scripps Research Institute, La Jolla, CA, USA; Shanghai Institute of Materia Medica, Chinese Academy of Sciences, Shanghai, China*

ROBERTO WURTH • *Department of Pharmacology and Physiology, Drexel University College of Medicine, Philadelphia, PA, USA; Section of Pharmacology, Department of Internal Medicine, University of Genoa, Genoa, Italy*

SIMON YONA • *The Rayne Institute, University College London, London, UK*

Chapter 1

The Fine Balance of Chemokines During Disease: Trafficking, Inflammation, and Homeostasis

Sandra M. Cardona, Jenny A. Garcia, and Astrid E. Cardona

Abstract

The action of chemokines (or "chemotactic cytokines") is recognized as an integral part of inflammatory and regulatory processes. Leukocyte mobilization during physiological conditions, trafficking of various cell types during pathological conditions, cell activation, and angiogenesis are among the target functions exerted by chemokines upon signaling via their specific receptors. Current research is focused in analyzing changes in chemokine/chemokine receptor patterns during various diseases with the aim to modulate pathological trafficking of cells, or to attract particular cell types to specific tissues. This review focuses on defining the role(s) of certain chemokine ligands and receptors in inflammatory neurological conditions such as multiple sclerosis. In addition, the role(s) of chemokines in neurodegenerative conditions such as Alzheimer's disease and Parkinson's disease is also described, as well as the contribution of chemokines to the pathogenesis of cancer, diabetes, and cardiovascular disease.

Key words Chemokines, Microglia, Monocytes, Multiple sclerosis, Diabetes, Cancer

1 Introduction

The chemokine family includes at least 46 human ligands that signal by binding to specific receptors. As implied by the name derived from *chemo*tactic cyto*kines*, these molecules are best known for their role in facilitating cellular migration. It is now recognized that chemokines are key players not only in inflammatory processes such as autoimmune disorders and infectious diseases but also in cancer, particularly in the regulation of metastasis. Chemokines are also important for normal cell positioning and pathfinding during development, cellular activation, and homeostatic cell trafficking under physiological conditions.

Most chemokines are small (8–14 KDa) proteins that interact with G-protein-coupled receptors. These molecules have been characterized predominantly in mammals but are also found in chicken and teleost fish [1], and possible homologs are also present in nematodes. There are 18 functionally signaling chemokine

Astrid E. Cardona and Eroboghene E. Ubogu (eds.), *Chemokines: Methods and Protocols*, Methods in Molecular Biology, vol. 1013, DOI 10.1007/978-1-62703-426-5_1, © Springer Science+Business Media New York 2013

receptors and four atypical chemokine receptors: Duffy antigen receptor for chemokines (DARC), D6, CCX-CKR, and CXCR7 which bind chemokines with high affinity but are unable to activate the signal transduction pathways that typical receptors do [2, 3]. Therefore, atypical chemokine receptors modulate immune responses by removing ligands from inflammatory sites; scavenging, sequestration, buffering, and intracellular transport best define their function [4].

Chemokines are divided into CC, CXC, CX3C, and XC groups according to their structure, arrangement, and spacing of two conserved cysteine residues. The CXC family has a non-conserved intervening amino acid residue in between the two cysteine residues, the CC family has two cysteine residues in juxtaposition, whereas in the CX3C family there are three intervening amino acids in between the two cysteine residues. The CXC family can be further subdivided into two categories based on the presence of the sequence glutamic acid–leucine–arginine (or ELR) at the NH_2 terminus immediately before the first cysteine of the CXC motif [5]. The ELR motif dictates chemokine specificity for binding to their cognate receptors. Similar to chemokines, chemokine receptors adopt a matching nomenclature, for example the chemokine fractalkine or CX3CL1 (L for ligand) binds to CX3CR1 (R for receptor).

Based on their expression patterns and functions, chemokines can also be grouped as follows: inflammatory (e.g., CCL1–13, CCL23, 24, CXCL1–3, CXCL5–11), i.e. involved in leukocyte mobilization during inflammation, or homeostatic (e.g., CCL14–16, CCL25, CCL27, CXCL12, 13) [6–8] implying involvement in leukocyte trafficking during normal physiological states. In addition, analyses of chemokine ligands and their receptor families show patterns of genomic organization attributable to evolutionary pressures [8]. Genomic organization further classifies chemokines into those whose genes are located in large clusters ("major-cluster" chemokines) and the "non-cluster" or "mini-cluster" chemokines whose genes are located separately in unique chromosomal locations [8]. In the human and mouse genomes, there are two major clusters of CC chemokine genes and two clusters of CXC genes, plus numerous non-clustered or mini-cluster genes of both types. Mouse models have provided significant knowledge of chemokine function under normal and pathophysiological conditions. However, as we extrapolate data from mice to humans, it is important to take into account the particular chemokine gene under study and the significant differences in the chemokine superfamily across mammalian species. We focus this review on the impact of chemokine biology and research on a chronic neuroinflammatory disorder such as multiple sclerosis (MS), neurodegenerative disorders such as Parkinson's disease (PD) and Alzheimer's disease (AD), and the pathology of diseases ranked as leading causes of death including cancer, diabetes, and cardiovascular disease.

2 Chemokines in Health and Disease

2.1 Chemokines and Central Nervous System Disorders

Under normal physiological conditions, the central nervous system (CNS) effectively controls and limits hematogenous leukocyte entry into the brain and spinal cord parenchyma. Multiple sclerosis is a disabling, neurological disorder that affects the CNS, causing loss of sensory and motor function. The etiology of MS is unknown. MS is considered a multifactorial or a polygenic disease, whereby multiple genes in combination with lifestyle and environmental factors contribute to disease development and progression. Inflammation, myelin loss, axonal degeneration, and progressive neurological dysfunction are the hallmarks of this disease [9–11].

In MS, allelic variation in the major histocompatibility complex (MHC) class II regions exerts the single strongest effect on genetic risk, especially those containing human leukocyte antigen (HLA)-DRB1*150 [12, 13]. Modestly powered genome-wide association studies have enabled the identification of more than 50 additional risk loci and multiple variants appear to exert modest individual effects in disease susceptibility [14]. Genes coding for cytokine pathways including the chemokine receptor CXCR5 and receptors for the cytokines interleukin (IL)-2, IL-7, IL-12, and tissue necrosis factor (TNF), costimulatory molecules (CD37, CD40, CD58, CD80, CD86, and CLECL1), signal transduction molecules (CBLB, GPR65, MALTI, RGS1, ATAT3, TAGAP, TYK2), the adhesion molecule vascular cell adhesion molecule (VCAM) among others [15] have been identified.

Sunlight, specifically through its role in producing active vitamin D, has been proposed as a key environmental factor for MS [16]. Studies of the vitamin D response elements (VDRE) and HLA-DRB1 haplotypes show that the most significant MHC-II risk haplotype for MS includes a functional VDRE in the proximal promoter region of HLA-DRB [17, 18]. The identified VDRE was found close to the highly conserved MHC-II-specific regulatory modules SXY important for constitutive and interferon (IFN)-γ-induced expression of HLA class II genes coordinated by the MHC2TA transactivator [18]. Although there is controversy whether MS is a primary autoimmune disorder, these genetic associations provide potential avenues to mechanistically determine relevant pathogenic interactions between susceptibility loci and environmental factors.

When considering how complex the inflammatory milieu is in MS with involvement of myeloid cells (e.g., monocytes, macrophages, dendritic cells, and microglia), T cells (e.g., CD4, CD8, γδ T cells, and effector Th1, Th17 subsets among others), B cells, natural killer (NK) cells, etc., the crucial role of chemokines in the timely trafficking of selected leukocyte subpopulations during disease progression requires emphasis. Blocking inflammatory cell influx to the CNS, in particular T cells (including the recently described subset of Th17 pathogenic T cells) and macrophages, has been a major focus for therapeutic intervention in MS. However, due to the fact that most leukocyte populations express more than one chemokine receptor and often one ligand can bind to more than one receptor, applying chemokine receptor blockade to ameliorate MS has been a challenging task.

Studies using the most widely accepted experimental animal model of MS, experimental autoimmune encephalomyelitis (EAE), reveal that blocking recruitment of myeloid cells (e.g., inflammatory Ly6C+CCR2+ monocytes) into the CNS is protective as shown by decreased EAE severity in mice lacking CCR2 [19, 20]. As for B cells, their accumulation is highly dependent on CXCL13 expression [21] and the implications of modulating B cell trafficking are currently unclear due to the fact that B cells can contribute to the pathogenesis of EAE through the production of anti-myelin antibodies that at least in part facilitate demyelination, as well as exert protective functions via regulation of effector T cell functions [22].

Several chemokine receptors mediate T cell trafficking. $CD4^+$ $CCR7^+$ memory T cells predominate in the CSF of MS patients. CXCR3 and CCR5 also play a role in T cell recruitment and expression of their chemokine ligands correlates with Th1 responses. Microglial activation also plays an important role in the pathology of MS and CX3CR1 was shown to inhibit microglial activation and neurotoxicity [23]. Together with CCR2, CX3CR1 not only controls the trafficking of selected monocyte populations during EAE but also exerts important regulatory roles on antigen-presenting cells and therefore impacts effector T cell responses [24]. Since blocking T cell and monocyte infiltration with chemokine receptor antagonists has been challenging thus far, chemokine receptors have also been used to target regulatory cells to specific organs. For instance, CCR5 has been utilized to home suppressive neural stem cells to the EAE brain and ameliorate inflammation [25].

In addition to chemoattraction, some chemokine receptors have been implicated in indirect modulation of effector function, such as CCR7 that acts on dendritic cells (DCs) to induce IL-23 production which in turn drives the generation of pathogenic Th17 cells [26]. Another chemokine with dual roles in the CNS is CXCL8; it may drive recruitment of neural stem cells and oligodendrocyte progenitors to sites of demyelination, but in turn could also contribute to myelin repair failure [27]. The complexity of the chemokine system in the CNS is further illustrated by CXCR4–CXCL12 interactions. CXCR4 is expressed by cells of the immune system and the CNS. Signaling in response to CXCL12 binding triggers migration and recruitment of immune cells including T cells and monocytes to the MS brain and its experimental models [28, 29], but is also important for the migration of neurons and oligodendrocyte precursor cells [30]. This ligand–receptor pair also plays a role in CNS development and organogenesis [31].

Alzheimer's and Parkinson's disease are the most common neurodegenerative disorders. AD accounts for about two-thirds of dementia cases, and inflammation is now recognized as an important component of its pathology. PD is characterized by the progressive loss of neurons in the substantia nigra, clinically manifesting with resting tremors, bradykinesia, rigidity, and postural instability. The genetic mapping gene-association studies as part of the Human Genome Project have accelerated the rate of identification of genes involved in AD and PD. It is accepted that aggregates of amyloid beta (Aβ)42 and α-synuclein are neurotoxic and contribute to AD and PD, respectively. However, chemokine expression profiles have provided critical information with reference to biomarker identification, as well as ways to accelerate aggregate clearance. For instance, CXCR4 expression preceded the loss of dopamine neurons in the 1-methyl-4-phenyl-1,2,3,6-tetrahydropyridine (MPTP)-model of PD, implying a possible molecular target for PD treatment [32]. Also, CX3CL1 was neuroprotective in the MPTP- and 6-hydroxydopamine (OHDA)-induced dopaminergic lesion models of PD [33, 34] via modulation of microglia activation.

Studies addressing the role of the CX3CL1 receptor, CX3CR1, in AD pathogenesis using CX3CR1-deficient mice have been somewhat discordant. The genetic model and analytical approaches used largely account for these differences. For instance, deleting CX3CR1 in tau transgenic mice exacerbated tau phosphorylation and aggregation as well as behavioral impairments [35]. In contrast, CX3CR1 deficiency prevented neuronal loss and microglial migration without affecting amyloid deposition in the triple transgenic model of AD harboring three mutant genes: presenilin-1 (PS1M146V), β-amyloid precursor protein (APPswe), and tau protein (TauP301L) [36]. Other studies showed that CX3CR1 deficiency attenuated amyloid deposition in AD models that develop high plaque deposition [37, 38]. A more recent study showed that CX3CR1 deficiency exacerbates functional deficits in human amyloid precursor protein (hAPP) transgenic mice, associated with elevated cytokine production and tau pathology and removing CX3CR1 in hAPP mice worsened memory retention [39].

Although serum chemokine analyses in PD patients did not reveal significant differences when compared with healthy controls, five CSF proteins (Aβ peptide 1–42, Flt3 ligand, CX3CL1, DJ1, and α-synuclein) were identified as potential biomarkers that may correlate with disease severity and progression [40]. Furthermore, early-stage AD patients with the highest levels of CSF CCL2 exhibited a significantly faster cognitive decline and developed dementia within a shorter time period compared to those with lower CCL2 levels. Therefore, CCL2 has been proposed as a candidate predictive biomarker of future AD progression [41].

2.2 Cancer

It is now widely recognized that tumor cells have the capability to modulate chemokine(s) and chemokine receptor(s) expression [42]. Chemokines can affect tumor spreading by promoting angiogenesis via interaction with their receptors on endothelial cells or indirectly via attracting inflammatory cells that release angiogenic factors such as fibroblast growth factor (bFGF) and vascular endothelial cell growth factor (VEGF) from the periphery to the tumor site. In addition, chemokines secreted by tumor cells have autocrine effects and induce tumor growth by interacting with chemokine receptors on tumor cells. Stromal cells within the tumor such as fibroblasts, DCs, lymphocytes, macrophages, and neutrophils also participate in the disease process, as they can become activated through direct cell–cell interactions, or by cytokines and chemokines they produce creating a positive feedback in tumor growth [43]. The CXC chemokine subfamily significantly contributes to the body's immune response to tumors, as well as tumor pathology. In vivo data indicate that certain chemokines can serve as tissue-specific attractant molecules for tumor cells. Particularly, ELR^+ CXC members are increased in a variety of tumors, and correlate with a more invasive (metastatic) and more vascularized (angiogenic) tumor phenotype. Therefore, the ELR^+ CXC chemokines can promote neovascularization, while ELR^- CXC chemokines can inhibit tumor angiogenesis acting on tumor growth and repression, respectively [5].

CXCR4 has also been implicated in the formation of metastases. CXCR4 is the chemokine receptor most commonly found in human and murine cancer cells, and has been implicated in the metastatic dissemination of tumors including pancreatic cancers, astrogliomas, myelomas, B cell lymphomas, and chronic lymphocytic leukemias [44]. Furthermore, CXCL12 and the two receptors it binds to, CXCR4 and CXCR7, are widely expressed in normal tissues and are involved in fetal development, mobilization of hematopoietic stem cells, and trafficking of naïve lymphocytes. CXCR4–CXCL12 interactions and downstream signaling, often partially regulated by Akt phosphorylation, promote tumor cell growth and survival allowing them to proliferate at distant and less favorable sites from the primary tumor. This phenomenon results from a mixed array of pathogenic responses such as increased cell migration, actin polymerization, and increased expression of angiogenic mediators such as VEGF. Chemokines and chemokine receptors may also exert indirect roles during tumor spreading; CXCL12 and CXCL10 act as activators of matrix metalloproteinase (MMP)-9 secretion, a molecule shown to be an important mediator of colorectal cancer invasion and metastasis [45].

Chemokines such as CXCL1, CXCL2, CXCL3, and CXCL8 have been studied in melanoma and shown to directly contribute to tumor cell growth, angiogenesis, and protection of tumor cells against apoptosis [46]. CXCL8 may facilitate tumor metastasis through its autocrine growth-promoting function as well as its induction of MMP activity, promoting the transmigration of tumor cells through basement membranes [47]. CXCL8 was also associated with increased invasiveness through the transcriptional activation of the MMP-2 gene, augmenting collagenase activity in human melanoma cells [48]. Similarly, in prostate cancer CXCL8 over-expression induced the expression of MMP-9 that in turn increased tumor cell invasiveness and metastatic potential in nude mice [49].

Inflammatory cytokines such as CXCL1 and CXCL8 can enhance tumor cell proliferation and have effects on angiogenesis. The production of these cytokines can be induced by the mutated human cancer oncogene *Ras* and oncogenic components of the EGFR–Ras–Raf signaling pathway [50, 51]. On the other hand, *Myc*, a gene involved in an array of cellular functions, is found over-expressed in many tumor cells where it maintains key aspects of the tumor phenotype including tumor cell autonomous proliferation and remodeling of the extracellular microenvironment. Moreover, studies in a pancreatic islet tumor model showed that Myc-activated cells produced chemokines with subsequent recruitment of mast cells capable of inducing angiogenesis and tumor growth [52, 53]. Based on the information above, chemokine gradients play their role by attracting tumor-promoting cells, but it is important to consider that such gradients can be modified by malignant cells in a manner that the absence of specific chemokines such as CXCL4 prevents the recruitment of circulating dendritic cell precursors and their in situ differentiation, which results in an advantageous environment for the tumor cells [54].

There is evidence that interactions between chemokine receptors expressed in cancer cells and the corresponding chemokines secreted within the target organs mediate organ-specific metastasis [44]. Chemokines expressed by specific organs promote tumor cell adhesion to microvessel walls, facilitating extravasation into target tissues and leading to tumor cell migration [55]. CXCR4 has been implicated in the organ-specific metastatic spreading of colorectal cancer to the liver [56–60]. It was also reported that intestinal metastasis only occurs in patients whose original biopsies had CCR9+ malignant cells [61]. CX3CR1 is expressed in tumors of neural origin such as gliomas and neuroblastomas where it participates in adhesion, transendothelial migration, and mobilization of the malignant cells [62]. High levels of CX3CR1 are also expressed in pancreatic cancer where it is involved in invasion of cells along pancreatic nerves. Similarly, other types of epithelial cancer cells, such as breast and prostate, express high levels of CX3CR1 [63].

It is well established that certain chronic inflammatory diseases such as prostatitis, hepatitis, and pancreatitis predispose to cancer. Cancer development under these inflammatory conditions is thought to be related to a "built-in" chemokine microenvironment caused by tissue damage [64]. Therefore, the interplay between inflammation and cancer is supported by de novo production of inflammatory mediators and certain chemokines, as well as their receptors, which are downstream of an oncogenic change [65]. CC chemokines, particularly CCL2 and CCL5, may recruit macrophage precursors into the tumor microenvironment contributing to the local inflammatory reaction [65–68].

As described above, CXC chemokines are key players at multiple stages of tumor progression and metastasis. This favors the recognition of chemokines and their receptors and downstream signaling pathways for treating malignant tumors. Monoclonal antibodies against chemokine receptors and neutralizing monoclonal antibodies against chemokines have been used successfully to inhibit the growth and migration of malignant tumor cells in experimental settings. Antibodies to CCL2 significantly reduced tumor burden in mice bearing prostate cancer [69] and reduced metastasis and prolonged survival in a mouse model of breast cancer [67]. Blockade of the CXCL12–CXCR4 signaling pathway can suppress cancer cell migration, invasion, angiogenesis, and metastases formation [70, 71]. The use of a newly developed specific inhibitor of the CXCR4 receptor (plerixafor) was successful in mobilizing hematopoietic stem cells to peripheral blood and is FDA approved for use in adult patients with non-Hodgkin lymphoma or multiple myeloma [72, 73]. Anti-angiogenic properties of CXCL9 and CXCL10 have been used as therapeutic controllers of tumor growth. Administration of CXCL10 or CXCL9 has shown to decrease tumor growth and angiogenesis in mice with adenocarcinomas [74] and non-small-cell lung cancer [75]. The therapeutic modulation of chemokine ligand–receptor signaling in cancers needs to be approached with extreme caution as chemokines have pivotal roles in healthy tissues.

2.3 Diabetes and Cardiovascular Disease

According to the American Diabetes Association (ADA), there are approximately 25.8 million children and adults diagnosed with diabetes throughout the United States of America, accounting for about 8.3 % of the population. Type I diabetes mellitus (T1DM), formerly known as juvenile diabetes, prevails in children and young adults. T1DM is classified as an immune-mediated disease secondary to the destruction of beta cells of the pancreas responsible for insulin production, resulting in hyperglycemia. Type II diabetes mellitus (T2DM) is the most common form of diabetes, accounting for about 90 % of cases and is highly associated with obesity. In T2DM either the body does not produce enough insulin or target organs are insensitive to it. Common complications of diabetes include cardiovascular disease, stroke, diabetic retinopathy, nephropathy, and peripheral neuropathy.

Cytokine and chemokine expression in T2DM patients is associated with activation of the immune system and has conferred an important role to inflammation in its pathogenesis, either as a consequence of oxidative stress [76, 77], endoplasmic reticulum stress [78], pancreatic amyloid deposition [79], lipid deposition and lipotoxicity [80, 81], and glucotoxicity [82, 83] or as a consequence of all of the above pathologic processes [84–88]. Inflammatory mediators upregulated in the obese state are more significantly produced by adipose tissue macrophages as opposed to adipocytes [89, 90]. The observation that macrophage cell counts are increased in obese individuals [89] suggests a role for these cells in normal adipocyte function. The phenotype of adipose tissue macrophages remains to be studied, along with their specific role(s) in insulin resistance.

Components of the adaptive immune system are also found in adipose tissue in T2DM. These include T cells ($CD4^+$ and $CD8^+$), $CD4^+$ T helper cells (Th1 and Th2 phenotypes), along with regulatory T cells (Treg) [91], associated with elevated levels of cytokines and chemokines. CCL2 is found at higher levels in the adipose tissue and circulation of obese patients compared to lean controls [92, 93]. Importantly, CCL2 and CXCL10 levels were found significantly higher in epithelial membranes and vitreous samples of patients suffering from diabetic retinopathy, suggesting a role of these chemokines in the recruitment of macrophages and monocytes to the diabetic eye [94–97]. These findings are consistent with studies showing a decreased number of macrophages in the adipose tissue of mice deficient in CCL2 or its receptor, CCR2 [98]. In vitro studies revealed increased production of CXCL8 and CCL3 by islet cells treated with high concentrations of glucose [99]. Similarly, rodent models of T2DM showed an increased production of CXCL1, CCL2, and CCL3 chemokines from isolated islets [100].

Histopathological examination of the pancreas from diabetic organ donors showed significant upregulation of chemokines CCL5, CCL8, CCL22, CXCL9, CXCL10, and CX3CL1 [101]. In experimental models, CCR5 was shown to contribute to monocyte infiltration to diabetic retinas [102] and adipose tissue and CCR5-deficient bone marrow conferred protection against impaired glucose tolerance [103]. In contrast, CXCR3-deficient NOD mice exhibited an accelerated development of diabetes attributed to the lack of Tregs required to effectively migrate into and regulate Th1 T-cell mediated inflammation within the pancreatic islets [104].

Analyses of a large cohort of German Caucasians and studies from Turkey and China reported a decreased prevalence of T2DM in populations with the CCL2 G-2518 gene variant [105]. It was also reported that insulin infusion can exert an anti-inflammatory effect by suppressing plasma concentrations of key chemokines involved in monocyte trafficking including CCL2 and CCL5 and their respective receptors CCR2 and CCR5 [106]. Due to the importance of monocytes/macrophages in the pathology of diabetes, anti-mouse CCL2 drugs [107] and CCR2 antagonists [108] have been developed to treat T2DM diabetic nephropathy and other associated complications. The validation and applicability to T2DM patients are still uncertain. CX3CL1 was also identified at higher levels in the subcutaneous adipose tissues of obese diabetic patients when compared to lean patients and this ligand was shown to modulate monocyte adhesion to adipocytes. Therefore, CX3CL1 was associated with obesity, insulin resistance, and T2DM and has been suggested as a diagnostic and therapeutic target in cardiometabolic disease [109].

CX3CL1 is accepted by many in the field as an adipose-expressed chemokine that best links T2DM and one of its major complications, heart disease [110]. The association between polymorphisms in its receptor, CX3CR1, and heart disease has been a topic of interest over the last 10 years. In humans, two single-nucleotide polymorphisms give rise to four allelic receptor variants. The most studied forms are CX3CR1$^{V249/T280}$ (also considered as the wild-type receptor), and the variant alleles CX3CR1$^{I249/T280}$ and CX3CR1$^{I249/M280}$ present in 20–30 % of the population. These changes decrease fractalkine affinity and correlate with enhanced susceptibility to age-related macular degeneration [111, 112] and protection from atherosclerosis [113–115]. CX3CR1 genotyping was performed in a cohort of 339 Caucasian individuals who underwent cardiac catheterization. An association was observed between CX3CR1 I249 allele expression and reduced prevalence of coronary artery disease (CAD). Furthermore, endothelium-dependent vasodilation was greater in these individuals compared to individuals heterozygous for the CX3CR1-V249 allele [115].

Another study investigated the association of the CX3CR1 alleles M280 and I249 with CAD and with acute coronary syndrome (ACS). Although the study found no association between expression of the I249 or the M280 allele with the occurrence of CAD, I249 and M280 appeared to have opposite effects on the occurrence of ACS, with allele I249 conferring an elevated risk [116]. The protective or detrimental effect of the CX3CR1 polymorphic variants was also investigated in early-onset and late-onset CAD patients. The I249 mutation was associated with multiple coronary vessel disease in older patients while V249 homozygosity was associated with the early-onset CAD [114]. In the Ludwigshafen Risk and Cardiovascular Health cohort, no relationship was found between CX3CR1 genotypes or corresponding haplotypes and the prevalence of CAD or myocardial infarction [117]. Although the role of CX3CR1 human variants in heart disease is controversial, the role of CX3CL1 in heart disease is further supported by significantly increased serum levels in unstable angina pectoris patients with plaque rupture compared with patients in other groups.

3 Concluding Remarks

There has been significant recent interest in modulating the interactions between chemokines and their receptors as novel anti-inflammatory drugs. Chemokine receptors have proved difficult to antagonize, largely due to their promiscuity and the compensatory loops available within the chemokine subfamilies. Clinical trials for CCR1, CCR2, CCR3, CCR5, CCR9, CXCR1, CXCR2, CXCR3, and CXCR4 are ongoing [110, 118, 119] for treatment of arthritis, multiple sclerosis, T2DM, HIV, inflammatory bowel disease, chronic obstructive pulmonary disease, psoriasis, and cancer. Plerixafor (CXCR4 blocker) has been commercially available since 2008 and is proven to effectively mobilize hematopoietic stem cells in combination with the granulocyte colony-stimulating factor. Therefore, the common features of these G-protein-coupled receptors may continue providing avenues to modulate common chemokine-dependent second messenger signaling pathways. With the advanced technologies available and the rapid progress in our understanding of leukocyte trafficking to sites of inflammation and tissue, the interest to target-specific chemokine receptors still continues and efforts are geared towards developing efficacious chemokine receptor antagonists to treat neuroinflammatory disorders, as well as systemic and peripheral tissue-restricted inflammatory diseases.

Acknowledgements

We are indebted to Dr. Eroboghene Ubogu for his critical review of the manuscript and to Elizabeth Morris for assistance in preparing the manuscript. The Astrid E. Cardona Laboratory is funded by the US National Multiple Sclerosis Society Grant TA3021A1/T, The San Antonio Area Foundation grant 201135345, and the US National Institutes of Health grants SC1GM095426 and R01NS078501.

References

1. Alejo A, Tafalla C (2011) Chemokines in teleost fish species. Dev Comp Immunol 35: 1215–1222
2. Ulvmar MH, Hub E, Rot A (2011) Atypical chemokine receptors. Exp Cell Res 317: 556–568
3. Graham GJ, Locati M, Mantovani A, Rot A, Thelen M (2012) The biochemistry and biology of the atypical chemokine receptors. Immunol Lett 145:30–38
4. Hansell CA, Hurson CE, Nibbs RJ (2011) DARC and D6: silent partners in chemokine regulation? Immunol Cell Biol 89:197–206
5. Strieter RM, Polverini PJ, Kunkel SL, Arenberg DA, Burdick MD, Kasper J et al (1995) The functional role of the ELR motif in CXC chemokine-mediated angiogenesis. J Biol Chem 270:27348–27357
6. IUISW/WHO Subcommittee on Chemokine nomenclature (2003) Chemokine/chemokine receptor nomenclature. Cytokine 21, 48–49
7. Bacon K, Baggiolini M, Broxmeyer H, Horuk R, Lindley I, Mantovani A et al (2002) Chemokine/chemokine receptor nomenclature. J Interferon Cytokine Res 22: 1067–1068

8. Zlotnik A, Yoshie O, Nomiyama H (2006) The chemokine and chemokine receptor superfamilies and their molecular evolution. Genome Biol 7:243
9. Noseworthy JH, Lucchinetti C, Rodriguez M, Weinshenker BG (2000) Multiple sclerosis. N Engl J Med 343:938–952
10. Dutta R, Trapp BD (2011) Mechanisms of neuronal dysfunction and degeneration in multiple sclerosis. Prog Neurobiol 93:1–12
11. Trapp BD, Nave KA (2008) Multiple sclerosis: an immune or neurodegenerative disorder? Annu Rev Neurosci 31:247–269
12. Chao MJ, Barnardo MC, Lincoln MR, Ramagopalan SV, Herrera BM, Dyment DA et al (2008) HLA class I alleles tag HLA-DRB1*1501 haplotypes for differential risk in multiple sclerosis susceptibility. Proc Natl Acad Sci USA 105:13069–13074
13. Ramagopalan SV, Herrera BM, Bell JT, Dyment DA, Deluca GC, Lincoln MR et al (2008) Parental transmission of HLA-DRB1*15 in multiple sclerosis. Hum Genet 122:661–663
14. Baranzini SE, Nickles D (2012) Genetics of multiple sclerosis: swimming in an ocean of data. Curr Opin Neurol 25:239–245
15. Sawcer S, Hellenthal G, Pirinen M, Spencer CC, Patsopoulos NA, Moutsianas L et al (2011) Genetic risk and a primary role for cell-mediated immune mechanisms in multiple sclerosis. Nature 476:214–219
16. Acheson ED, Bachrach CA, and Wright FM. (1960) Some comments on the relationship of the distribution of multiple sclerosis to latitude, solar radiation, and other variables. Acta Psychiatr Scand Suppl 35:132–147
17. Ramagopalan SV, Heger A, Berlanga AJ, Maugeri NJ, Lincoln MR, Burrell A et al (2010) A ChIP-seq defined genome-wide map of vitamin D receptor binding: associations with disease and evolution. Genome Res 20:1352–1360
18. Ramagopalan SV, Maugeri NJ, Handunnetthi L, Lincoln MR, Orton SM, Dyment, DA et al (2009) Expression of the multiple sclerosis-associated MHC class II Allele HLA-DRB1*1501 is regulated by vitamin D. PLoS Genet 5, e1000369
19. Saederup,N., Cardona,A.E., Croft,K., Mizutani,M., Cotleur,A.C., Tsou,C.L.et al. (2010) Selective chemokine receptor usage by central nervous system myeloid cells in CCR2-red fluorescent protein knock-in mice. PLoS One 5, e13693-
20. Bennett JL, Elhofy A, Charo I, Miller SD, Dal Canto MC, Karpus WJ (2007) CCR2 regulates development of Theiler's murine encephalomyelitis virus-induced demyelinating disease. Viral Immunol 20:19–33
21. Kowarik MC, Cepok S, Sellner J, Grummel V, Weber MS, Korn T, et al (2012) CXCL13 is the major determinant for B cell recruitment to the CSF during neuroinflammation. J Neuroinflammation 9:93
22. Mann MK, Ray A, Basu S, Karp CL, Dittel BN (2012) Pathogenic and regulatory roles for B cells in experimental autoimmune encephalomyelitis. Autoimmunity 45: 388–399
23. Cardona A, Pioro EP, Sasse ME, Kostenko V, Cardona SM, Dijkstra IM et al (2006) Control of microglial neurotoxicity by the fractalkine receptor. Nat Neurosci 9:917–924
24. King IL, Dickendesher TL, Segal BM (2009) Circulating Ly-6C+ myeloid precursors migrate to the CNS and play a pathogenic role during autoimmune demyelinating disease. Blood 113:3190–3197
25. Yang J, Yan Y, Ma CG, Kang T, Zhang N, Gran B et al (2012) Accelerated and enhanced effect of CCR5-transduced bone marrow neural stem cells on autoimmune encephalomyelitis. Acta Neuropathol
26. Kuwabara T, Ishikawa F, Yasuda T, Aritomi K, Nakano H, Tanaka Y et al (2009) CCR7 ligands are required for development of experimental autoimmune encephalomyelitis through generating IL-23-dependent Th17 cells. J Immunol 183:2513–2521
27. Kelland EE, Gilmore W, Weiner LP, Lund BT (2011) The dual role of CXCL8 in human CNS stem cell function: Multipotent neural stem cell death and oligodendrocyte progenitor cell chemotaxis. Glia 59:1864–1878
28. Meiron M, Zohar Y, Anunu R, Wildbaum G, Karin N (2008) CXCL12 (SDF-1alpha) suppresses ongoing experimental autoimmune encephalomyelitis by selecting antigen-specific regulatory T cells. J Exp Med 205: 2643–2655
29. Kohler RE, Comerford I, Townley S, Haylock-Jacobs S, Clark-Lewis I, McColl SR (2008) Antagonism of the chemokine receptors CXCR3 and CXCR4 reduces the pathology of experimental autoimmune encephalomyelitis. Brain Pathol 18:504–516
30. Banisadr G, Frederick TJ, Freitag C, Ren D, Jung H, Miller SD et al (2011) The role of CXCR4 signaling in the migration of transplanted oligodendrocyte progenitors into the cerebral white matter. Neurobiol Dis 44: 19–27
31. Zou YR, Kottmann AH, Kuroda M, Taniuchi I, Littman DR (1998) Function of the chemokine receptor CXCR4 in haematopoiesis

and in cerebellar development. Nature 393:595–599
32. Shimoji M, Pagan F, Healton EB, Mocchetti I (2009) CXCR4 and CXCL12 expression is increased in the nigro-striatal system of Parkinson's disease. Neurotox Res 16: 318–328
33. Rogers JT, Morganti JM, Bachstetter AD, Hudson CE, Peters MM, Grimmig BA et al (2011) CX3CR1 deficiency leads to impairment of hippocampal cognitive function and synaptic plasticity. J Neurosci 31: 16241–16250
34. Pabon MM, Bachstetter AD, Hudson CE, Gemma C, Bickford PC (2011) CX3CL1 reduces neurotoxicity and microglial activation in a rat model of Parkinson's disease. J Neuroinflammation 8:9
35. Bhaskar K, Konerth M, Kokiko-Cochran ON, Cardona A, Ransohoff RM, Lamb BT (2010) Regulation of tau pathology by the microglial fractalkine receptor. Neuron 68:19–31
36. Fuhrmann M, Bittner T, Jung CK, Burgold S, Page RM, Mitteregger G et al (2010) Microglial Cx3cr1 knockout prevents neuron loss in a mouse model of Alzheimer's disease. Nat Neurosci 13:411–413
37. Lee S, Varvel NH, Konerth ME, Xu G, Cardona AE, Ransohoff RM et al (2010) CX3CR1 deficiency alters microglial activation and reduces beta-amyloid deposition in two Alzheimer's disease mouse models. Am J Pathol 177:2549–2562
38. Liu Z, Condello C, Schain A, Harb R, Grutzendler J (2010) CX3CR1 in microglia regulates brain amyloid deposition through selective protofibrillar amyloid-beta phagocytosis. J Neurosci 30:17091–17101
39. Cho SH, Sun B, Zhou Y, Kauppinen TM, Halabisky B, Wes P et al (2011) CX3CR1 protein signaling modulates microglial activation and protects against plaque-independent cognitive deficits in a mouse model of Alzheimer disease. J Biol Chem 286:32713–32722
40. Scalzo P, de Miranda AS, Guerra Amaral DC, de CV, Cardoso F, Teixeira AL (2011) Serum levels of chemokines in Parkinson's disease. Neuroimmunol 18:240–244
41. Westin K, Buchhave P, Nielsen H, Minthon L, Janciauskiene S, Hansson O (2012) CCL2 is associated with a faster rate of cognitive decline during early stages of Alzheimer's disease. PLoS One 7, e30525
42. Arya M, Patel HR, Williamson M (2003) Chemokines: key players in cancer. Curr Med Res Opin 19:557–564
43. Zhu,Q., Han,X., Peng,J., Qin,H., and Wang,Y. (2012) The role of CXC chemokines and their receptors in the progression and treatment of tumors. J Mol Histol 43(6):699–713
44. Balkwill F (2004) The significance of cancer cell expression of the chemokine receptor CXCR4. Semin Cancer Biol 14:171–179
45. Zipin-Roitman A, Meshel T, Sagi-Assif O, Shalmon B, Avivi C, Pfeffer RM et al (2007) CXCL10 promotes invasion-related properties in human colorectal carcinoma cells. Cancer Res 67:3396–3405
46. Mehrad B, Keane MP, Strieter RM (2007) Chemokines as mediators of angiogenesis. Thromb Haemost 97:755–762
47. Singh RK, Gutman M, Radinsky R, Bucana CD, Fidler IJ (1994) Expression of interleukin 8 correlates with the metastatic potential of human melanoma cells in nude mice. Cancer Res 54:3242–3247
48. Youngs SJ, Ali SA, Taub DD, Rees RC (1997) Chemokines induce migrational responses in human breast carcinoma cell lines. Int J Cancer 71:257–266
49. Inoue K, Slaton JW, Kim SJ, Perrotte P, Eve BY, Bar-Eli M et al (2000) Interleukin 8 expression regulates tumorigenicity and metastasis in human bladder cancer. Cancer Res 60:2290–2299
50. Guerra C, Schuhmacher AJ, Canamero M, Grippo PJ, Verdaguer L, Perez-Gallego L et al (2007) Chronic pancreatitis is essential for induction of pancreatic ductal adenocarcinoma by K-Ras oncogenes in adult mice. Cancer Cell 11:291–302
51. Sparmann A, Bar-Sagi D (2004) Ras-induced interleukin-8 expression plays a critical role in tumor growth and angiogenesis. Cancer Cell 6:447–458
52. Soucek L, Lawlor ER, Soto D, Shchors K, Swigart LB, Evan GI (2007) Mast cells are required for angiogenesis and macroscopic expansion of Myc-induced pancreatic islet tumors. Nat Med 13:1211–1218
53. Sodir NM, Swigart LB, Karnezis AN, Hanahan D, Evan GI, Soucek L (2011) Endogenous Myc maintains the tumor microenvironment. Genes Dev 25:907–916
54. Song EY, Shurin MR, Tourkova IL, Gutkin DW, Shurin GV (2010) Epigenetic mechanisms of promigratory chemokine CXCL14 regulation in human prostate cancer cells. Cancer Res 70:4394–4401
55. Erez N, Coussens LM (2011) Leukocytes as paracrine regulators of metastasis and determinants of organ-specific colonization. Int J Cancer 128:2536–2544
56. Gunther K, Leier J, Henning G, Dimmler A, Weissbach R, Hohenberger W et al (2005) Prediction of lymph node metastasis in

colorectal carcinoma by expressionof chemokine receptor CCR7. Int J Cancer 116: 726–733
57. Kim J, Mori T, Chen SL, Amersi FF, Martinez SR, Kuo C et al (2006) Chemokine receptor CXCR4 expression in patients with melanoma and colorectal cancer liver metastases and the association with disease outcome. Ann Surg 244:113–120
58. Kim J, Takeuchi H, Lam ST, Turner RR, Wang HJ, Kuo C et al (2005) Chemokine receptor CXCR4 expression in colorectal cancer patients increases the risk for recurrence and for poor survival. J Clin Oncol 23: 2744–2753
59. Zeelenberg IS, Ruuls-Van SL, Roos E (2003) The chemokine receptor CXCR4 is required for outgrowth of colon carcinoma micrometastases. Cancer Res 63:3833–3839
60. Ottaiano,A., di,P.A., Napolitano,M., Pisano,C., Pignata,S., Tatangelo,F.et al. (2005) Inhibitory effects of anti-CXCR4 antibodies on human colon cancer cells. Cancer Immunol Immunother 54, 781–791.
61. Amersi FF, Terando AM, Goto Y, Scolyer RA, Thompson JF, Tran AN et al (2008) Activation of CCR9/CCL25 in cutaneous melanoma mediates preferential metastasis to the small intestine. Clin Cancer Res 14:638–645
62. Marchesi F, Locatelli M, Solinas G, Erreni M, Allavena P, Mantovani A (2010) Role of CX3CR1/CX3CL1 axis in primary and secondary involvement of the nervous system by cancer. J Neuroimmunol 224:39–44
63. Marchesi F, Piemonti L, Mantovani A, Allavena P (2010) Molecular mechanisms of perineural invasion, a forgotten pathway of dissemination and metastasis. Cytokine Growth Factor Rev 21:77–82
64. Mantovani A, Allavena P, Sica A, Balkwill F (2008) Cancer-related inflammation. Nature 454:436–444
65. Allavena P, Germano G, Marchesi F, Mantovani A (2011) Chemokines in cancer related inflammation. Exp Cell Res 317:664–673
66. Vetrano S, Borroni EM, Sarukhan A, Savino B, Bonecchi R, Correale C et al (2010) The lymphatic system controls intestinal inflammation and inflammation-associated Colon Cancer through the chemokine decoy receptor D6. Gut 59:197–206
67. Qian BZ, Li J, Zhang H, Kitamura T, Zhang J, Campion LR et al (2011) CCL2 recruits inflammatory monocytes to facilitate breast-tumour metastasis. Nature 475:222–225
68. Bottazzi B, Polentarutti N, Acero R, Balsari A, Boraschi D, Ghezzi P et al (1983) Regulation of the macrophage content of neoplasms by chemoattractants. Science 220: 210–212
69. Zhang J, Patel L, Pienta KJ (2010) CC chemokine ligand 2 (CCL2) promotes prostate cancer tumorigenesis and metastasis. Cytokine Growth Factor Rev 21:41–48
70. Arya M, Ahmed H, Silhi N, Williamson M, Patel HR (2007) Clinical importance and therapeutic implications of the pivotal CXCL12-CXCR4 (chemokine ligand-receptor) interaction in cancer cell migration. Tumour Biol 28:123–131
71. Wang J, Loberg R, Taichman RS (2006) The pivotal role of CXCL12 (SDF-1)/CXCR4 axis in bone metastasis. Cancer Metastasis Rev 25:573–587
72. DiPersio JF, Micallef IN, Stiff PJ, Bolwell BJ, Maziarz RT, Jacobsen E et al (2009) Phase III prospective randomized double-blind placebo-controlled trial of plerixafor plus granulocyte colony-stimulating factor compared with placebo plus granulocyte colony-stimulating factor for autologous stem-cell mobilization and transplantation for patients with non-Hodgkin's lymphoma. J Clin Oncol 27:4767–4773
73. DiPersio JF, Stadtmauer EA, Nademanee A, Micallef IN, Stiff PJ, Kaufman JL et al (2009) Plerixafor and G-CSF versus placebo and G-CSF to mobilize hematopoietic stem cells for autologous stem cell transplantation in patients with multiple myeloma. Blood 113:5720–5726
74. Dias S, Choy M, Rafii S (2001) The role of CXC chemokines in the regulation of tumor angiogenesis. Cancer Invest 19:732–738
75. Addison CL, Arenberg DA, Morris SB, Xue YY, Burdick MD, Mulligan MS et al (2000) The CXC chemokine, monokine induced by interferon-gamma, inhibits non-small cell lung carcinoma tumor growth and metastasis. Hum Gene Ther 11:247–261
76. Evans JL, Goldfine ID, Maddux BA, Grodsky GM (2002) Oxidative stress and stress-activated signaling pathways: a unifying hypothesis of type 2 diabetes. Endocr Rev 23: 599–622
77. Evans JL, Goldfine ID, Maddux BA, Grodsky GM (2003) Are oxidative stress-activated signaling pathways mediators of insulin resistance and beta-cell dysfunction? Diabetes 52:1–8
78. Hotamisligil GS (2010) Endoplasmic reticulum stress and the inflammatory basis of metabolic disease. Cell 140:900–917
79. Zraika S, Hull RL, Verchere CB, Clark A, Potter KJ, Fraser PE et al (2010) Toxic oligomers and islet beta cell death: guilty by association or convicted by circumstantial evidence? Diabetologia 53:1046–1056

80. Unger RH (1995) Lipotoxicity in the pathogenesis of obesity-dependent NIDDM. Genetic and clinical implications Diabetes 44: 863–870
81. Unger RH, Clark GO, Scherer PE, Orci L (2010) Lipid homeostasis, lipotoxicity and the metabolic syndrome. Biochim Biophys Acta 1801:209–214
82. Yki-Jarvinen H, Helve E, Koivisto VA (1987) Hyperglycemia decreases glucose uptake in type I diabetes. Diabetes 36:892–896
83. Rossetti L, Smith D, Shulman GI, Papachristou D, DeFronzo RA (1987) Correction of hyperglycemia with phlorizin normalizes tissue sensitivity to insulin in diabetic rats. J Clin Invest 79:1510–1515
84. Hotamisligil GS, Erbay E (2008) Nutrient sensing and inflammation in metabolic diseases. Nat Rev Immunol 8:923–934
85. Donath MY, Storling J, Maedler K, Mandrup-Poulsen T (2003) Inflammatory mediators and islet beta-cell failure: a link between type 1 and type 2 diabetes. J Mol Med (Berl) 81: 455–470
86. Ehses JA, Ellingsgaard H, Boni-Schnetzler M, Donath MY (2009) Pancreatic islet inflammation in type 2 diabetes: from alpha and beta cell compensation to dysfunction. Arch Physiol Biochem 115:240–247
87. Donath MY, Schumann DM, Faulenbach M, Ellingsgaard H, Perren A, Ehses JA (2008) Islet inflammation in type 2 diabetes: from metabolic stress to therapy. Diabetes Care 31(Suppl 2):S161–S164
88. Masters SL, Dunne A, Subramanian SL, Hull RL, Tannahill GM, Sharp FA et al (2010) Activation of the NLRP3 inflammasome by islet amyloid polypeptide provides a mechanism for enhanced IL-1beta in type 2 diabetes. Nat Immunol 11:897–904
89. Weisberg SP, McCann D, Desai M, Rosenbaum M, Leibel RL, Ferrante AW Jr (2003) Obesity is associated with macrophage accumulation in adipose tissue. J Clin Invest 112: 1796–1808
90. Xu H, Barnes GT, Yang Q, Tan G, Yang D, Chou CJ et al (2003) Chronic inflammation in fat plays a crucial role in the development of obesity-related insulin resistance. J Clin Invest 112:1821–1830
91. Wu H, Ghosh S, Perrard XD, Feng L, Garcia GE, Perrard JL et al (2007) T-cell accumulation and regulated on activation, normal T cell expressed and secreted upregulation in adipose tissue in obesity. Circulation 115:1029–1038
92. Breslin WL, Johnston CA, Strohacker K, Carpenter KC, Davidson TR, Moreno JP et al (2012) Obese Mexican American children have elevated MCP-1, TNF-alpha, monocyte concentration, and dyslipidemia. Pediatrics 129:e1180–e1186
93. Catalan V, Gomez-Ambrosi J, Ramirez B, Rotellar F, Pastor C, Silva C et al (2007) Proinflammatory cytokines in obesity: impact of type 2 diabetes mellitus and gastric bypass. Obes Surg 17:1464–1474
94. Abu El-Asrar AM, Struyf S, Kangave D, Geboes K, Van DJ (2006) Chemokines in proliferative diabetic retinopathy and proliferative vitreoretinopathy. Eur Cytokine Netw 17:155–165
95. Harada C, Okumura A, Namekata K, Nakamura K, Mitamura Y, Ohguro H et al (2006) Role of monocyte chemotactic protein-1 and nuclear factor kappa B in the pathogenesis of proliferative diabetic retinopathy. Diabetes Res Clin Pract 74:249–256
96. Hernandez C, Segura RM, Fonollosa A, Carrasco E, Francisco G, Simo R (2005) Interleukin-8, monocyte chemoattractant protein-1 and IL-10 in the vitreous fluid of patients with proliferative diabetic retinopathy. Diabet Med 22:719–722
97. Capeans C, De Rojas MV, Lojo S, Salorio MS (1998) C-C chemokines in the vitreous of patients with proliferative vitreoretinopathy and proliferative diabetic retinopathy. Retina 18:546–550
98. Kanda H, Tateya S, Tamori Y, Kotani K, Hiasa K, Kitazawa R et al (2006) MCP-1 contributes to macrophage infiltration into adipose tissue, insulin resistance, and hepatic steatosis in obesity. J Clin Invest 116:1494–1505
99. Ehses JA, Perren A, Eppler E, Ribaux P, Pospisilik JA, Maor-Cahn R et al (2007) Increased number of islet-associated macrophages in type 2 diabetes. Diabetes 56: 2356–2370
100. Ehses JA, Lacraz G, Giroix MH, Schmidlin F, Coulaud J, Kassis N et al (2009) IL-1 antagonism reduces hyperglycemia and tissue inflammation in the type 2 diabetic GK rat. Proc Natl Acad Sci USA 106:13998–14003
101. Sarkar SA, Lee CE, Victorino F, Nguyen TT, Walters JA, Burrack A et al (2012) Expression and regulation of chemokines in murine and human type 1 diabetes. Diabetes 61: 436–446
102. Serra AM, Waddell J, Manivannan A, Xu H, Cotter M, Forrester JV (2012) CD11b(+) Bone Marrow-Derived Monocytes Are the Major Leukocyte Subset Responsible for Retinal Capillary Leukostasis in Experimental Diabetes in Mouse and Express High Levels of CCR5 in the Circulation. Am J Pathol 181: 719–727

103. Kitade H, Sawamoto K, Nagashimada M, Inoue H, Yamamoto Y, Sai Y et al (2012) CCR5 plays a critical role in obesity-induced adipose tissue inflammation and insulin resistance by regulating both macrophage recruitment and M1/M2 status. Diabetes 61:1680–1690
104. Yamada Y, Okubo Y, Shimada A, Oikawa Y, Yamada S, Narumi S et al (2012) Acceleration of diabetes development in CXC chemokine receptor 3 (CXCR3)-deficient NOD mice. Diabetologia 55:2238–2245
105. Panee J (2012) Monocyte Chemoattractant Protein 1 (MCP-1) in obesity and diabetes. Cytokine 60:1–12
106. Ghanim H, Korzeniewski K, Sia CL, Abuaysheh S, Lohano T, Chaudhuri A et al (2010) Suppressive effect of insulin infusion on chemokines and chemokine receptors. Diabetes Care 33:1103–1108
107. Ninichuk V, Clauss S, Kulkarni O, Schmid H, Segerer S, Radomska E et al (2008) Late onset of Ccl2 blockade with the Spiegelmer mNOX-E36-3'PEG prevents glomerulosclerosis and improves glomerular filtration rate in db/db mice. Am J Pathol 172:628–637
108. Sullivan TJ, Dairaghi DJ, Krasinski A, Miao Z, Wang Y, Zhao BN et al (2012) Characterization of CCX140-B, an orally bioavailable antagonist of the CCR2 chemokine receptor, for the treatment of type 2 diabetes and associated complications. J Pharmacol Exp Ther
109. Shah R, Hinkle CC, Ferguson JF, Mehta NN, Li M, Qu L et al (2011) Fractalkine is a novel human adipochemokine associated with type 2 diabetes. Diabetes 60:1512–1518
110. Charo IF, Ransohoff RM (2006) The many roles of chemokines and chemokine receptors in inflammation. N Engl J Med 354:610–21
111. Tuo J, Smith BC, Bojanowski CM, Meleth AD, Gery I, Csaky KG et al (2004) The involvement of sequence variation and expression of CX3CR1 in the pathogenesis of age-related macular degeneration. FASEB J 18:1297–1299
112. Chan CC, Tuo J, Bojanowski CM, Csaky KG, Green WR (2005) Detection of CX3CR1 single nucleotide polymorphism and expression on archived eyes with age-related macular degeneration. Histol Histopathol 20: 857–863
113. Moatti D, Faure S, Fumeron F, Amara M, Seknadji P, McDermott DH et al (2001) Polymorphism in the fractalkine receptor CX3CR1 as a genetic risk factor for coronary artery disease. Blood 97:1925–1928
114. Nassar BA, Nanji AA, Ransom TP, Rockwood K, Kirkland SA, Macpherson K et al (2008) Role of the fractalkine receptor CX3CR1 polymorphisms V249I and T280M as risk factors for early-onset coronary artery disease in patients with no classic risk factors. Scand J Clin Lab Invest 68:286–291
115. McDermott DH, Halcox JP, Schenke WH, Waclawiw MA, Merrell MN, Epstein N et al (2001) Association between polymorphism in the chemokine receptor CX3CR1 and coronary vascular endothelial dysfunction and atherosclerosis. Circ Res 89:401–407
116. Niessner A, Marculescu R, Haschemi A, Endler G, Zorn G, Weyand CM et al (2005) Opposite effects of CX3CR1 receptor polymorphisms V249I and T280M on the development of acute coronary syndrome. A possible implication of fractalkine in inflammatory activation. Thromb Haemost 93:949–954
117. Matzhold EM, Trummer O, Grunbacher G, Zulus B, Boehm BO, Marz W et al (2009) Association of polymorphisms in the chemokine receptor CX3CR1 gene with coronary artery disease. Cytokine 47:224–227
118. Horuk R (2009) Chemokine receptor antagonists: overcoming developmental hurdles. Nat Rev Drug Discov 8:23–33
119. Horuk R, Proudfoot AE (2009) Drug discovery targeting the chemokine system–where are we? Front Biosci (Elite Ed) 1:209–219

Chapter 2

Chemokine-Dependent Signaling Pathways in the Peripheral Nervous System

Eroboghene E. Ubogu

Abstract

Chemokines and their G-protein-coupled receptors play important roles in development, homeostasis, and the innate and adaptive immune response. Pathologic chemokine signaling pathways in the peripheral nervous system can be studied in peripheral nerves using human in vitro models of the blood–nerve barrier (BNB) and a reliable model of acute peripheral nerve inflammation called severe murine experimental autoimmune neuritis (EAN). This chapter describes a flow-dependent human leukocyte-BNB trafficking assay and the reliable induction of EAN in female SJL/J mice as tools to study pro-inflammatory chemokine-dependent signaling in peripheral nerves.

Key words Blood–nerve barrier, Chemokines, Chemokine receptors, Endoneurial endothelial cells, Experimental autoimmune neuritis, Guillain–Barré syndrome, Leukocyte trafficking, Peripheral nerves

1 Introduction

Autoimmune neuropathies (AIN) can be classified as a group of restricted peripheral nervous system (PNS) disorders in which the presumed pathogenic mechanism is immune-mediated attack on components of the peripheral nerves, nerve roots, or both. These may be simplistically divided into Guillain–Barré syndrome (GBS), chronic inflammatory demyelinating polyradiculoneuropathy (CIDP), and vasculitic neuropathy (VN). These disorders are characterized by hematogenous leukocyte infiltration (predominantly monocytes and T-cells) into peripheral nerves, with resultant demyelination and axonal degeneration. Inflammation also plays a central role in the response to nerve injury, with the activation of resident immune cells and infiltration of hematogenous leukocytes (neutrophils, monocytes, T-cells). These cells help remove debris and facilitate repair. Chemokine-mediated signaling has been implicated in the pathogenesis of AIN and its animal

Astrid E. Cardona and Eroboghene E. Ubogu (eds.), *Chemokines: Methods and Protocols*, Methods in Molecular Biology, vol. 1013, DOI 10.1007/978-1-62703-426-5_2, © Springer Science+Business Media New York 2013

models, as well as the pathophysiology of neuropathic pain based on rodent pain hypersensitivity models [1–7].

Due to the separation of peripheral nerves from the systemic immune system by the BNB, mechanisms of leukocyte extravasation into the endoneurium (innermost compartment of nerves) across the BNB are pathologically relevant in GBS. Human endoneurial endothelial cells (that form the BNB) express an array of chemokines under basal conditions and following pro-inflammatory cytokine treatment in vitro [8]. The sequential process of chemokine-dependent leukocyte trafficking across the BNB can be studied in real time using a flow-dependent leukocyte trafficking assay that involves a flow chamber coupled to time-lapse video microscopy [8–12].

Experimental autoimmune neuritis (EAN) is considered a standard animal model for the most common demyelinating form of GBS called acute inflammatory demyelinating polyradiculoneuropathy. EAN can be induced through immunization of animals with heterogeneous peripheral nerve myelin or its protein components such as P2 or P0 emulsified in complete Freund's adjuvant (CFA) [13]. Severemurine EAN (sm-EAN) is a robust GBS animal model due to very high induction rates, similarity to human demyelinating GBS, and strong correlation between electrophysiologic and pathologic characteristics with disease severity, compared with other murine EAN models [14]. In sm-EAN, 8–12-week-old female SJL/J mice are immunized with bovine peripheral nerve myelin (BPNM), using pertussis toxin and recombinant mouse interleukin-12 (IL-12) as co-adjuvants [14–16].

2 Materials

2.1 Flow-Dependent Leukocyte-BNB Assay

1. Primary human endoneurial endothelial cells (pHEndECs, *see* **Note 1**).
2. pHEndEC medium: RPMI-1640 medium with phenol red, 10 % NuSerum, 10 % fetal bovine serum, 1× penicillin–streptomycin, 1× vitamin solution, 1× nonessential amino acid solution, 2 mM l-glutamine, 1 nM sodium pyruvate, 10 mM HEPES buffer, 1 ng/mL recombinant human basic fibroblast growth factor (hbFGF), 10 U/mL heparin sodium salt, 50 μg/mL endothelial cell growth supplements (ECGS). Stock solutions of RPMI-1640 medium with phenol red, sodium pyruvate, nonessential amino acid solution, and 1 M HEPES, as well as heparin sodium salt, are stored at 4 °C, while stock solutions or aliquots of NuSerum, fetal bovine serum, penicillin–streptomycin, vitamin solution, 200 mM glutamine, hbFGF, and ECGS are kept at −20 °C (*see* **Note 2**). Thaw frozen components in a 37 °C water bath, and add all components into a medium bottle,

filtering across a 0.2 μm membrane using a presterilized vacuum filter disposable unit. Add RPMI-1640 medium with phenol red last. Adjust pH to 7.4, and store at 4 °C (*see* **Note 3**).

3. Rat tail collagen (RTC) solution: 50 μg/mL Type I RTC, 0.02 N acetic acid in sterile deionized distilled water (ddH_2O). Stock solution of Type I RTC (formulated at variable concentrations in 0.02 N acetic acid) is kept at 4 °C, while 2 N acetic acid solution in ddH_2O and ddH_2O are kept at room temperature (*see* **Note 4**). Filter components as in previous step, adding ddH_2O last. Store at 4 °C under sterile conditions.
4. 35 mm Coring CellBIND® tissue culture dishes (*see* **Note 5**).
5. Non-cytotoxic round 14 mm diameter silicone well dividers: Transparent CultureWell™ silicone well dividers (Electron Microscopy Services, Hartfield, PA, USA) are recommended. Cut to fit 35 mm tissue culture dishes, wash in 70 % ethanol, and then 5 % sodium hypochlorite solution (bleach). Rinse thoroughly with water. Wrap in aluminum foil and sterilize by autoclaving prior to use (*see* **Note 6**).
6. Ammonium hydroxide chamber: 5 mL ammonium hydroxide (ammonia content 28–30 %, kept at 4 °C) poured into a 100 mm diameter non-tissue culture-treated Petri dish and put in a covered Styrofoam or plastic box placed in a biosafety cabinet (*see* **Note 7**).
7. Peripheral blood mononuclear leukocytes (PBMLs): Isolated from fresh whole heparinized blood using Ficoll® density centrifugation (*see* **Note 8**).
8. Assay buffer: 1 % bovine serum albumin (fraction V) in RPMI-1640 medium without phenol red (both kept at 4 °C). Weigh 5.0 g and prepare a 500 mL solution by gently pouring albumin into a flask with RPMI-1640 medium, and mix by gentle spinning using a magnetic stirrer (*see* **Note 9**). Filter as in **step 2** and aliquot into 50 mL centrifuge tubes and store at 4 °C. Warm and maintain at 37 °C prior to use.
9. Parallel plate flow chamber kit for 35 mm tissue culture dishes (e.g., GlycoTech, Gaithersburg, MD, USA): Cast acrylic flow deck, silicon rubber gasket, polypropylene tubing connectors, Silastic™ laboratory tubing [8–12].
10. Infusion/withdrawal pumps, e.g., KD scientific model 210 pump (Holliston, MA, USA). Two pumps are needed (*see* **Note 10**).
11. 10 and 30 mL Luer-Lok™ tip syringes.
12. Video microscopy system: Inverted phase contrast microscope with heated stage, digital camera, and data processing software (e.g., Zeiss Axiovert 40 Inverted Microscope, Zeiss Axiocam

MRc 5 digital camera with Axiovision software, National Institutes of Health Image J software).

13. Agonists and inhibitors: Human recombinant cytokines and chemokines, mouse anti-human chemokine receptor antibodies, small molecular antagonists.

2.2 Severe Murine Experimental Autoimmune Neuritis

1. 8–12-week-old female SJL/J mice, weighing 20–25 g (*see* **Note 11**).
2. BPNM: Isolated from the cauda equina of healthy young adult United States Department of Agriculture-certified cows <30 months of age, as previously published [17] (*see* **Note 12**). Lyophilized BPNM is stored under anhydrous conditions in uncovered 50 mL centrifuge tubes placed in a sealed plastic container partly filled with size 4 mesh indicating Drierite at −20 °C.
3. Adjuvants and co-adjuvants: CFA (containing 1 mg/mL *Mycobacterium tuberculosis* H37RA), pertussis toxin, recombinant mouse interleukin-12 (rm-IL12). Store CFA at room temperature and shake thoroughly before use. Make 100 μg/mL pertussis toxin in 1× phosphate-buffered saline (PBS), aliquot 50 μL into Eppendorf tubes, and store at 4 °C. Make 100 μg/mL rm-IL12 in 1 % BSA dissolved in 1× PBS (*see* Subheading 2.1 item 8, substituting RPMI-1640 medium with 1× PBS), aliquot 50 μL into Eppendorf tubes, and store at −20 °C (*see* **Note 13**).
4. 5 or 10 mL Luer-Lok™ tip syringes.
5. 3-way stopcocks, e.g., Discofix® 3-way stopcock (B. Braun Medical, Inc., Bethlehem, PA, USA).
6. 1 mL Tuberculin syringes with sterile slip tip 26 G and permanent 28 G needles (*see* **Note 14**).
7. 18 G single-use sterile needles.
8. Anesthesia solution: Ketamine hydrochloride, xylazine hydrochloride. Make 8 mg/mL ketamine solution in 1× PBS and store at 4 °C. Make 0.8 mg/mL xylazine solution in 1× PBS and store at −20 °C. Ketamine is a restricted, controlled substance, so it should be stored and its use recorded according to local, state, or federal regulations.
9. Exogenous agonists/antagonists: For example chemokine receptor inhibitors (function-neutralizing monoclonal antibodies, small molecular inhibitors).

3 Methods

3.1 Flow-Dependent Leukocyte-BNB Assay

1. Plate 10^6 cryopreserved pHEndECs in a 75 cm^2 RTC-coated CellBIND® tissue culture flask, culture in a humidified incubator with 95 % air, 5 % CO_2, and harvest with 0.05 % trypsin–0.53 mM

EDTA in Hank's balanced salt solution when >80 % confluent, as previously published [18, 19] (*see* **Note 15**). Prepare 35 mm tissue culture dishes prior to harvesting pHEndECs.

2. Place silicone well dividers in the center of the 35 mm tissue culture dish with sterilized forceps soaked in 70 % ethanol. Number plates to facilitate easy identification (*see* **Note 16**).
3. Add 100 μL of the RTC solution into the center of the well divider, gently swirling to ensure complete coating of the dish base.
4. Carefully place uncovered 35 mm tissue culture dishes by hand in the covered ammonium hydroxide chamber kept in a biosafety cabinet for 10 min to precipitate the RTC and form a gel coating.
5. Remove plates from the ammonium hydroxide chamber and carefully add 100 μL of 1× PBS to the RTC gel-solution in each plate, and incubate covered for 60 min at room temperature in the biosafety cabinet. Carefully and completely aspirate solution from the edge of the well divider, and add 100 μL of 1× PBS. Keep until ready to plate the cells (*see* **Note 17**).
6. Harvest pHEndECs with 0.05 % typsin–0.53 mM EDTA as previously published [20], quantify, and resuspend at 10^6 cells/mL in pHEndEC medium warmed to 37 °C.
7. Completely aspirate 1× PBS from the edge of the silicone well divider and add 100 μL warmed pHEndEC medium in the center of each dish. Swirl to evenly distribute.
8. Carefully triturate pHEndECs and add 50 μL (50,000 cells) into the center of each dish. Cover and allow cells to adhere to the dish for 30 min at room temperature in the biosafety cabinet (*see* **Note 18**). Gently add 2 mL of warmed pHEndEC medium over the silicone well divider to ensure complete coverage, and place in the incubator. Remove silicone well dividers and change medium after 24–48 h (*see* **Note 19**). Observe endothelial cell growth and layer formation daily with a phase contrast microscope and change medium (2 mL pre-warmed pHEndEC medium per dish) every 48 h. Dishes are ready for use after 5–7 days in culture, forming an in vitro BNB.
9. The in vitro BNB may be treated with pro-inflammatory cytokines (e.g., 10 U/mL TNF-α + 20 U/mL IFN-γ) for up to 48 h prior to conducting trafficking assays, as previously published [8] (*see* **Note 20**).
10. Wash fresh or cryopreserved PBMLs (10^7 cells in 1 mL) by adding to 9 mL of warmed assay buffer in a centrifuge tube and spin at 300 × *g* for 5 min at room temperature in a benchtop centrifuge. Completely aspirate supernatant, resuspend in 1 mL assay buffer, aliquot 200 μL (2×10^6 cells) into 1.5 mL Eppendorf tubes, and store on ice at 4 °C (*see* **Note 21**).

11. Set up camera software to take time-lapsed, high-resolution grayscale digital images (e.g., Axiocam MRc5 black/white camera mode, 1,298×968 pixels/in. test standard mono) every second. Set up phase contrast microscope, and visualize using the 20–40× objective lens (*see* **Note 22**). Set heated stage to 37 °C and verify steady uniform temperature with a thermometer (e.g., infrared laser digital thermometer). Clean all surfaces with 70 % ethanol.
12. Calculate flow rates (in mL/min) to generate a constant linear velocity of 1–2 mm/s (estimated red blood cell capillary flow velocity) based on the parallel plate flow chamber gasket dimensions and manufacturer's instructions. Set up infusion and withdrawal pumps, paying attention to the make and volume of the syringes used. The 10 mL syringe should be used for infusion, with the 30 mL syringe used for withdrawal.
13. Set up parallel plate flow chamber according to the manufacturer's instructions. Clean all surfaces and tubes with 70 % ethanol followed by 1× PBS before use. Ensure that the gasket is firmly attached to the flow deck, and outflow tube to the withdrawal pump (*see* **Note 23**). Attach a 10 mL syringe with 1× PBS to the inflow tube and gently infuse solution to eliminate air bubbles within the tube. Keep the PBS in the tube until ready to perform the assay.
14. Take an aliquot of PBMLs (from **step 10**), wash by adding 800 μL warmed assay buffer into the Eppendorf tube, and spin at 9,000×*g* for 30 s at room temperature on a laboratory benchtop microcentrifuge. Resuspend in 1 mL assay buffer with or without inhibitors (e.g., chemokine receptor antagonists) and incubate for 5–10 min at 37 °C. Aspirate into a 10 mL syringe containing 3 mL assay buffer (final concentration 500,000 cells/mL). Triturate several times using a sterile large-bore (e.g., 18 G) needle to achieve a uniform cell suspension and eliminate all air bubbles (*see* **Note 24**). Withdraw the PBS from the inflow tube, attach syringe, and manually infuse PBML suspension to the inlet of the flow chamber deck, expelling all air bubbles. Attach syringe to the infusion pump.
15. In experiments requiring BNB pretreatment with agonists (e.g., chemokines) or inhibitors (e.g., antibodies against adhesion molecules, small molecular signaling pathway antagonists), the tissue culture dish should be washed with assay buffer and treated for 5–10 min with the agonist/antagonist at 37 °C in the humidified incubator. This should be timed to facilitate the commencement of the trafficking assay within 1–2 min of preparing the PBML suspension. Remove the treated or untreated 35 mm tissue culture dish from the incubator and carefully dump medium into a waste beaker, leaving a small film of medium over the surface to prevent the endothelial cell layer from drying out.

16. Firmly attach the flow deck to the 35 mm dish, taking care not to excessively move the inflow tube and generate air bubbles. Ensure that the deck is flush with the dish surface. Apply medium-sized binder clips (e.g., 5/8 in. capacity) to the outer part of the flow chamber just before the inlet and outlet to ensure firm contact (*see* **Note 25**).
17. Place the flow chamber system on the microscope stage, focus on the endothelial cell layer, adjust brightness, and start infusion at a higher rate to ensure that medium completely covers the field of view within a few seconds. Adjust infusion and withdrawal pumps to the required flow rates and start taking time-lapse digital images for 20–30 min. Once the leukocyte trafficking assay is complete, pause the infusion and withdrawal pumps, detach 35 mm dish from flow chamber deck, aspirate residual PBMLs from the inlet tube, and clean the deck and gasket with 70 % ethanol using delicate task wipes. Repeat **steps 13–17** depending on the experimental design.
18. Merge sequential digital images (Fig. 1) from a single experiment to generate a video microscopy data file for subsequent analysis using appropriate software or manually. Data from the same experiment should be saved in a single folder for easier processing (*see* **Note 26**).

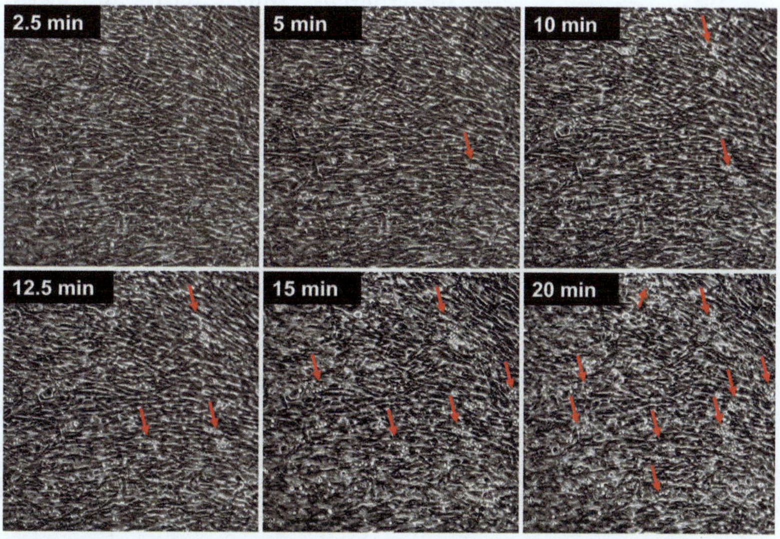

Fig. 1 Leukocyte-BNB trafficking assay. PBMLs from an untreated GBS patient were infused at 1 mm/s across the in vitro BNB treated with 10 U/mL TNF-α + 20 U/mL IFN-γ for 24 h. Digital phase contrast photomicrographs were obtained by automated time-lapse image acquisition. Numbers at the *upper left corner* depict time points after initiating PBML infusion. Clusters of firmly adherent PBMLs (*arrows*) become noticeable after about 5 min, with progressive increase in the number and size of PBML clusters with time, reaching a peak between 20 and 25 min

3.2 Experimental Autoimmune Neuritis

1. Weigh lyophilized BPNM in an analytical balance by placing into a pre-weighed 1.5 mL Eppendorf tube with a clean spatula. Each mouse requires 2.5 mg of BPNM. Weigh at least 25 % more BPNM than required to account for losses during emulsification. Add sterile 1× PBS to the Eppendorf tube to make a 25 mg/mL suspension, and keep on ice at 4 °C (*see* **Note 27**).
2. Using an 18 G needle attached to the Luer Lok™ syringe, mix several times and completely aspirate BPNM–PBS suspension. If initial suspension concentration was >25 mg/mL, add the additional amount of 1× PBS into the Eppendorf tube. Mix and collect using the previously used needle and syringe. If volume is <2 mL, the 5 mL syringe is preferred.
3. Add an equal volume of thoroughly mixed CFA to the Eppendorf tube (no more than 1 mL at a time), mix, and collect with the same syringe and needle used in **step 2**. This ensures complete BPNM transfer prior to emulsification, with a final concentration of 12.5 mg/mL.
4. Emulsify BPNM–PBS–CFA mix by tightly connecting the syringe to one of the ports of a sterile 3-way stopcock. Attach an empty syringe of equal capacity to the other syringe port. Mix 500 times on ice to completely emulsify the BPNM. Keep emulsified BPNM on ice prior to use (*see* **Note 28**).
5. Anesthetize mice with pre-warmed 100 mg/kg ketamine + 10 mg/kg xylazine in the same sterile 1 mL tuberculin syringe with the permanent 28 G needle via intraperitoneal injection. Other anesthetic agents (e.g., halothane) may be used to maintain anesthesia for at least 30 min. Clean the backs of mice with 70 % ethanol when there is minimal response to tail pinch. Tag mice to allow for easy identification.
6. Remove the empty syringe from its stopcock port. Take off the 26 G sterile slip tip needle from a 1 mL tuberculin syringe and attach a sterile 18 G needle. Carefully withdraw no more than 600 μL BPNM into syringe by placing needle into the empty port and slowly plunging syringe with the emulsified BPNM to release it into the port (*see* **Note 29**). Remove 18 G needle and replace with 26 G needle.
7. Slowly inject 50 μL of emulsified BPNM subcutaneously at four sites (tail base, left flank, right flank, and mid back) on the mouse's back, a total of 200 μL per mouse. Pinch the skin upwards and introduce the needle parallel to the back's surface to prevent intramuscular injection (*see* **Note 30**). After each injection, pinch the puncture site for at least 10 s to prevent BPNM from leaking out. Change 26 G needles between mice.
8. Repeat **steps 5–7** depending on the number of mice being studied. Place mice prone on a heating blanket. Place in cages

with sufficient bedding once early signs of arousal are noted. Observe every 6–12 h and give scheduled analgesia (e.g., buprenorphine 1 mg/kg) by intraperitoneal injection for the next 24–48 h.

9. Make a fresh 2 μg/mL solution of pertussis toxin in 1× PBS and inject 100 μL (200 ng) intraperitoneally into each mouse on the day of induction (day 0) using a sterile 1 mL tuberculin syringe. Inject the same amount of pertussis toxin on day 2 post induction (*see* **Note 31**).
10. Make a fresh 1 μg/mL solution of rm-IL12 in 1× PBS and inject 100 μL (100 ng) intraperitoneally into each mouse on days 1, 2, and 3 post induction as above (*see* **Note 32**).
11. Perform daily weights and neuromuscular severity scores, as published [14, 15, 21]. Administer chemokine signaling agonists or antagonists at predefined intervals, depending on the experimental design. A video clip of a female SJL/J mouse with severe EAN at expected peak severity is accessible at the following link: http://www.bcm.edu/neurology/neuromuscularlab/index.cfm?PMID=24242 (as of March, 2013).

4 Notes

1. pHEndECs are not currently available commercially. Detailed protocols for isolating these cells from the sciatic nerves of recently decedent individuals and verifying phenotypic characteristics have been published [20].
2. Aliquot components to the volumes needed to make medium prior to storage to prevent repetitive freeze–thaw cycles that may degrade protein components. Make heparin sodium salt solution in RPMI-1640 medium just before adding to make pHEndEC medium. Take care to note the weight to unit conversion for the heparin sodium salt stock. Use an analytical balance to measure out the exact weight of heparin in an Eppendorf tube, and then add 1 mL of RPMI-1640 medium to mix.
3. Total medium volume of 500 mL is recommended, depending on the frequency of use. pH 7.4 supports pHEndEC growth. Medium is good for repetitive use (warming to 37 °C prior to use) for about a month when stored and kept sterile at 4 °C.
4. The RTC stock concentration should be noted and volume adjusted to produce the required final concentration. 2 N acetic acid in ddH_2O can be made using fresh stock of >99 % glacial acetic acid and kept at room temperature for several months.

5. The Corning CellBIND® tissue culture-treated dishes support more rapid attachment, as well as superior proliferation and maturation of cultured pHEndECs compared to other tissue culture-treated dishes based on our experience.
6. Round silicone well dividers are preferred to square-shaped dividers, as these are better at concentrating pHEndEC growth in the middle of the tissue culture dish. Repetitive cleaning and sterilization of the recommended silicone wells do not affect performance.
7. Ammonium hydroxide is corrosive and has a strong pungent smell. Wear a mask and handle with care. Discard residual ammonium hydroxide in a dedicated waste bottle kept in a fume hood and dispose compound and Petri dish according to local waste regulations.
8. Controlled rate cryopreservation methods can be used to store isolated PBMLs in liquid nitrogen with excellent viability (>95 %) for several years. This is more practical when working with leukocytes from untreated patients with peripheral nerve inflammation due to unpredictable donor availability.
9. Heating or very rapid spinning could denature bovine serum albumin. It is important to allow albumin slowly dissolve over 15–30 min at room temperature.
10. Infusion mode is required for the flow chamber inlet, with withdrawal mode needed for outlet. Single-mode pumps can be used at either the inlet or the outlet.
11. Female SJL/J mice (from The Jackson Laboratory) are reliably susceptible to severe EAN induction using this protocol. Using age-matched littermates generated from in-house breeding pairs is preferable to recently purchased adult mice. Male mice and other female mouse strains demonstrate variable induction rates and varying degrees of severity. Susceptibility of female SJL/J mice <8 weeks or >12 weeks of age has not been evaluated.
12. BPNM is not currently available commercially. Bovine cauda equina can be purchased fresh from a local abattoir or frozen from an animal tissue or an organ product company such as Innovative Research (Novi, MI, USA). Cauda equina tissue may be stored at −80 °C until ready for use. It is crucial to completely dissect the cauda equina away from any residual spinal cord tissue prior to isolation to prevent BPNM contamination with central nervous system myelin.
13. Handle adjuvants and co-adjuvants with extreme care and avoid skin contact due to potential to induce inflammation. CFA precipitates on standing, necessitating thorough mixing prior to use. Large vials (e.g., 10 mL) may be aliquoted into

1.5 mL Eppendorf tubes prior to storage and used for several years. Pertussis toxin can be fatal if ingested. Stock solutions of pertussis toxin and rm-IL12 remain potent for at least 12 months. Working solutions should be made just before use, with residuals discarded.

14. Tuberculin syringes with slip tip 26 G needles are needed for subcutaneous BPNM injection while syringes with permanent 28 G needle are used for intraperitoneal injections. The 26 G needle is needed due to the viscosity of the emulsified BPNM.
15. Recently passaged pHEndECs are preferred to plating with cryopreserved cells due to better attachment, more uniform endothelial layer formation following culture, and more reliable expression of cellular adhesion molecules. Cells are usually ready for passaging 4–5 days after initial seeding.
16. Numbering helps limit errors with later wash, endothelial cell plating, and treatment steps.
17. Following the final 1× PBS wash, 200 μL pHEndEC medium may be placed in the center of the tissue culture dish (within the well divider) that can be sealed in Parafilm® and stored overnight at 4 °C. This is not recommended due to potential evaporation of medium owing to its small volume. It is advisable to plate pHEndECs on the same day the dishes are prepared.
18. Cell density and amount of cells seeded provide the most uniform endothelial cell layer in our experience. Maintaining dishes in the biosafety cabinet for 30 min facilitates uniform adherence to the RTC gel coat in the CellBIND® dishes.
19. The silicone well dividers are relatively hydrophobic, so sufficient volume of medium is needed to completely cover the surface of the dividers and its center. Do not add medium directly into the center to prevent detachment of seeded pHEndECs. The well dividers may detach from the dish surface after several hours. It is important to ensure that the dividers do not float over plated pHEndECs as this could affect gaseous exchange and induce cell death. Keeping well dividers in the dishes for more than 48 h could compromise uniform endothelial cell growth and cause cell layer clumping at its margins.
20. pHEndECs are serum dependent for anchorage, so it is advisable to include at least 90 % pHEndEC medium in the final solution containing cytokines or chemokines. We commonly use assay buffer (1 % BSA in RPMI-1640 medium without phenol red) to initially dilute cytokine stock.
21. Completely thaw vial of cryopreserved PBMLs for 1–2 min in a 37 °C water bath, gently aspirate the cell suspension, and

inject into pre-warmed assay buffer. No less than 1:5 dilution is recommended to wash cryopreserved cells so as to remove preservatives. Cell viability may be compromised by keeping cells in the water bath for too long, using smaller centrifugation volumes and higher centrifugation forces. Storing on ice prior to the trafficking assay limits PBML chemokine receptor internalization.

22. Note the lapse time between images and adjust settings accordingly if needed. It is highly recommended to visualize confluent pHEndEC layers prior to initial microscope setup to determine the required magnification needed for the trafficking assay based on the equipment available.

23. A small amount of vacuum grease provides a tight seal between the flow deck and the rubber gasket. Placing a few layers of Parafilm® around the edges of the flow deck will keep rubber gasket in place and ensure a tight fit with the 35 mm Corning CellBIND® tissue culture dishes. Avoid putting too much Parafilm®, as this would make removing tissue culture dishes difficult on completion of the trafficking assay.

24. Final leukocyte concentration and assay volumes can be modified as needed, aiming to maintain near-physiological concentrations in vivo and sufficient number of quantifiable trafficking cells. At least 1.5 mL extra PBML suspension is advised to account for non-visualized cells in the inlet tube and provide enough cells for the assay in the event of any technical challenges experienced while setting up the flow chamber. Air bubbles in the inflow tube and flow chamber adversely affect linear flow and can cause endothelial layer shearing. Air in the outflow tube does not affect the assay. The amount of inhibitor added should be calculated based on the effective dose per 10^6 cells.

25. This works very well with linear flow velocities <7 mm/s in this model system and allows for air bubble expulsion from the flow chamber prior to initiating the trafficking assay if needed. Vacuum connection is commonly used when studying leukocyte trafficking with higher flow velocities or shear stresses with macrovascular endothelial cells such as human umbilical vein endothelial cells [11, 12].

26. A computer with a dedicated large storage hard drive (e.g., 400 GB capacity), external hard drive, or access to institutional data storage drive is advised due to large high-resolution digital file size. 300–500 MB may be needed to store images from a 20–30-min study. We import the raw JPEG digital images to the NIH Image J software, compress to 50 % (reduces final video size by a factor of 4), and merge at 10 frames per second (FPS). A typical 20-min assay generates a 2-min video of

~368 MB size. It is prudent to store digital files and delete raw data from the acquisition software program after performing several assays to prevent compromising subsequent time-lapse image acquisition. We recommend maintaining no more than 8 GB worth of data (~16 independent 30-min trafficking assays) prior to hard drive storage. Trafficking events can be quantified manually or with cell tracking software.

27. BPNM contains >70 % lipid, so it is quite hydrophobic and sticky. There is no need to transfer BPNM–PBS suspension into a larger centrifuge tube. No more than 1 mL PBS should be used, implying that a higher stock BPNM–PBS concentration is acceptable. The final concentration should be 25 mg/mL prior to emulsification (*see* Subheading 3.2, **step 2**).
28. Emulsified BPNM in CFA is quite viscous and heat is generated during the process, necessitating the use of ice to prevent protein degradation. Improper emulsification results in unreliable disease induction. Position the syringe plunger in the most comfortable position for the operator. Taking breaks after 100–200 mixes should be considered.
29. Directly plunging BPNM from the Luer Lok™ into 1 mL tuberculin syringe may result in tuberculin syringe disconnection and BPNM spillage due to the myelin's viscosity and pressure buildup within the syringe during emulsification. Taking out no more than 600 mL (sufficient for three mice) facilitates easier removal of air bubbles and subcutaneous injection.
30. Intramuscular injection can result in muscle necrosis. Intravenous or intraperitoneal BPNM injection could result in death of the mouse, so these should not be performed. Skin ulceration may occur if the subcutaneous injection is too superficial.
31. Pertussis toxin is needed to increase microvascular permeability required to induce an inflammatory response. EAN is restricted to peripheral nerves due to antigenic stimulation by BPNM.
32. rm-IL12 is needed to polarize the BPNM-induced immune response towards a strong T-helper 1 phenotype (as observed in human GBS).

Acknowledgements

This work was initially supported by a Baylor College of Medicine New Investigator Start-Up Award and grants from the Guillain–Barré syndrome/Chronic Inflammatory Demyelinating Polyradiculoneuropathy Foundation International, and is currently supported by National Institutes of Health/National Institute of Neurological Disorders and Stroke grants NS073702, NS075212, and NS078226 to EEU.

References

1. Ubogu EE (2011) Chemokine receptors as specific anti-inflammatory targets in peripheral nerves. Endocr Metab Immune Disord Drug Targets 11:141–153
2. Kieseier B, Krivacic K, Jung S et al (2000) Sequential expression of chemokines in experimental autoimmune neuritis. J Neuroimmunol 110:121–129
3. Kinter J, Broglio L, Steck AJ et al (2010) Gene expression profiling in nerve biopsy of vasculitic neuropathy. J Neuroimmunol 225: 184–189
4. Orlikowski D, Chazaud B, Plonquet A et al (2003) Monocyte chemoattractant protein 1 and chemokine receptor CCR2 productions in Guillain-Barré syndrome and experimental autoimmune neuritis. J Neuroimmunol 134: 118–127
5. Perrin F, Lacroix S, Avilés-Trigueros M et al (2005) Involvement of monocyte chemoattractant protein-1, macrophage inflammatory protein-1alpha and interleukin-1beta in Wallerian degeneration. Brain 128:854–866
6. Press R, Pashenkov M, Jin J et al (2003) Aberrated levels of cerebrospinal fluid chemokines in Guillain-Barré syndrome and chronic inflammatory demyelinating polyradiculoneuropathy. J Clin Immunol 23:259–267
7. White F, Jung H, Miller R (2007) Chemokines and the pathophysiology of neuropathic pain. Proc Natl Acad Sci USA 104:20151–20158
8. Yosef N, Ubogu EE (2012) alpha(M) beta(2) -integrin-intercellular adhesion molecule-1 interactions drive the flow-dependent trafficking of Guillain-Barre syndrome patient derived mononuclear leukocytes at the blood-nerve barrier in vitro. J Cell Physiol 227:3857–3875. doi:10.1002/jcp. 24100
9. Alon R, Ley K (2008) Cells on the run: shear-regulated integrin activation in leukocyte rolling and arrest on endothelial cells. Curr Opin Cell Biol 20:525–532
10. Man S, Tucky B, Bagheri N et al (2009) alpha4 Integrin/FN-CS1 mediated leukocyte adhesion to brain microvascular endothelial cells under flow conditions. J Neuroimmunol 210:92–99
11. Shulman Z, Alon R (2009) Chapter 14. Real-time in vitro assays for studying the role of chemokines in lymphocyte transendothelial migration under physiologic flow conditions. Methods Enzymol 461:311–332
12. Gopalan P, Jones D, McIntire L et al (2001) Cell adhesion under hydrodynamic flow conditions. Curr Protoc Immunol Chapter 7: Unit 7.29
13. Meyer zu Hörste G, Hartung H, Kieseier B (2007) From bench to bedside—experimental rationale for immune-specific therapies in the inflamed peripheral nerve. Nat Clin Pract Neurol 3:198–211
14. Xia R, Yosef N, Ubogu E (2010) Clinical, electrophysiological and pathologic correlations in a severe murine experimental autoimmune neuritis model of Guillain-Barré syndrome. J Neuroimmunol 219:54–63
15. Xia RH, Yosef N, Ubogu EE (2010) Selective expression and cellular localization of pro-inflammatory chemokine ligand/receptor pairs in the sciatic nerves of a severe murine experimental autoimmune neuritis model of Guillain-Barré syndrome. Neuropathol Appl Neurobiol 36:388–398
16. Calida D, Kremlev S, Fujioka T et al (2000) Experimental allergic neuritis in the SJL/J mouse: induction of severe and reproducible disease with bovine peripheral nerve myelin and pertussis toxin with or without interleukin-12. J Neuroimmunol 107:1–7
17. Xia RH, Yosef N, Burns AR, Eichberg J, Ubogu EE (2012) Isolation, purification and verification of peripheral nerve myelin derived from bovine cauda equina. J Neurol Neurophysiol doi:10.4172/2155-9562.S7-002
18. Mahad D, Callahan M, Williams K et al (2006) Modulating CCR2 and CCL2 at the blood–brain barrier: relevance for multiple sclerosis pathogenesis. Brain 129:212–223
19. Ubogu E, Callahan M, Tucky B et al (2006) Determinants of CCL5-driven mononuclear cell migration across the blood–brain barrier. Implications for therapeutically modulating neuroinflammation. J Neuroimmunol 179: 132–144
20. Yosef N, Xia R, Ubogu E (2010) Development and characterization of a novel human in vitro blood-nerve barrier model using primary endoneurial endothelial cells. J Neuropathol Exp Neurol 69:82–97
21. Ubogu EE, Yosef N, Xia RH et al (2012) Behavioral, electrophysiological, and histopathological characterization of a severe murine chronic demyelinating polyneuritis model. J Peripher Nerv Syst 17:53–61

Chapter 3

Multispectral Imaging and Automated Laser Capture Microdissection of Human Cortical Neurons: A Quantitative Study of CXCR4 Expression

Jonathan Pitcher, Roberto Wurth, Saori Shimizu, and Olimpia Meucci

Abstract

Quantifying protein and RNA expression within specific cell populations in vivo is an essential step in unraveling the complex mechanisms of neurological disease. The challenges associated with studying human brain tissue are commonly compounded by variations in postmortem interval, formalin fixation time, and tissue processing methods among others. The result is a sample population that is inherently heterogeneous, implying the need for reliable protocols that are sensitive to low levels of antigen while minimizing background and nonspecific staining. Here, we describe a single immunohistochemistry protocol on formalin-fixed, paraffin-embedded human cortex which can be adapted to (1) quantify the relative protein expression of the chemokine receptor, CXCR4, using multispectral image or (2) isolate neuronal RNA through automated laser capture microdissection.

Key words Chemokine, CXCR4, Multispectral imaging, Immunohistochemistry, Human cortex, Neuron, Quantitative, LCM, Laser capture microdissection

1 Introduction

The protocol described is designed to measure protein and RNA expression of the chemokine receptor, CXCR4, using multispectral imaging and automated laser capture microdissection (LCM), respectively. Both protocols are based on an immunohistochemistry (IHC) protocol designed for optimal antigen retrieval and minimal nonspecific staining among 20 human tissue samples stained and analyzed within a single batch. Because of the variability inherent to any studied human population, as well as introduced by differing fixation protocols, particular attention is taken to thoroughly block all sources of nonspecific staining and cross-reaction between antibodies and respective chromogens [1–3]. Depending on the specific samples used, incubation and wash times may be altered without affecting staining quality. The following protocol should be optimized according to the specific tissue and antibodies used.

Astrid E. Cardona and Eroboghene E. Ubogu (eds.), *Chemokines: Methods and Protocols*, Methods in Molecular Biology, vol. 1013, DOI 10.1007/978-1-62703-426-5_3, © Springer Science+Business Media New York 2013

2 Materials

2.1 Reagents

All solutions, unless otherwise specified, are prepared using purified deionized water with a resistivity of 18 MΩ cm and <5 ppb total oxidizable carbon (TOC) at 25 °C. Phosphate-buffered saline (PBS) is a commonly used diluent in IHC, however, because phosphate-containing buffers (like PBS) should not be used with alkaline phosphatase-conjugated antibodies; Tris-buffered saline (TBS) is used instead. Therefore, all uses of TBS refer to the working concentration of TBS, pH 7.4, described in IHC reagents. Many of the reagents used, particularly the chromogens, are toxic and must be handled and disposed of properly. All IHC techniques should be carried out in a well-ventilated chemical fume hood, unless otherwise specified.

2.2 Equipment and Software

1. Nuance® multispectral imaging system with Nuance® version 2.1 software (CRi, Woburn, MA, USA).
2. Zeiss MicroBeam IV with PALM RoboSoftware version 4.2 Pro SP2 and AxioVision Release 4.8.1 (Carl Zeiss Microscopy, Thornwood, NY, USA).

2.3 Immunohistochemistry

1. Positive charged slides, Coplin jars, and slide rack.
2. Shandon xylene substitute (Thermo-Scientific) (*see* **Note 1**).
3. Ethyl alcohol (EtOH), 200 proof (*see* **Note 2**).
4. Target retrieval solution, 10× concentrate, (Dako).
5. PAP hydrophobic barrier pen, (VWR).
6. Avidin/Biotin blocking kit (Vector Labs).
7. Normal donkey serum (NDS) (Jackson ImmunoResearch) (*see* **Note 3**).
8. Mouse monoclonal anti-MAP2A/B (microtubule associated protein 2), clone AP20, IgG1, (Chemicon/Millipore).
9. Mouse monoclonal anti-CXCR4, clone 12 G5, $IgG2_A$ (R&D Systems).
10. Biotin-SP (long spacer, 6-atom spaced position between the antibody and the biotin)-conjugated AffiniPure Donkey Anti-mouse IgG (H+L) secondary antibody (Jackson ImmunoResearch).
11. Avidin biotin complex (ABC) amplification solution horseradish peroxidase (HRP) (Vector Labs).
12. Avidin biotin complex amplification solution alkaline phosphatase, ABC-AP (Vector Labs).
13. Levamisole (Vector Labs).
14. Vector® NovaRED™ substrate kit for peroxidase (Vector Labs) (*see* **Note 4**).

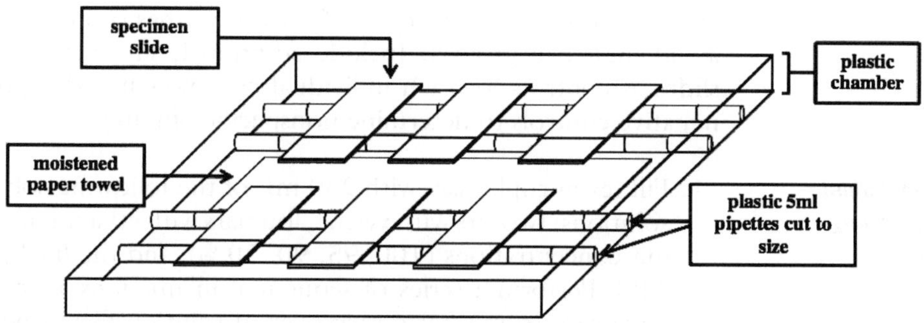

Fig. 1 Self-assembled humidity chamber. A humidity chamber can be assembled using a flat, shallow plastic tray with a lid (lid not shown), four plastic pipette tips, and a paper towel. Simply cut the pipette tips to a length that allows them to fit snugly within the tray in two pairs. Place a paper towel on the bottom of the tray and saturate with Millipore water and then place slides, face up, on the pipette tips. During incubations, cover the plastic tray with a fitted top and for overnight incubations seal the top with a band of Parafilm around the perimeter of the chamber

15. Vector® Blue alkaline phosphatase substrate kit III (Vector Labs).
16. VectaMount™ permanent mounting medium (Vector Labs).
17. TBS pH 7.4 (10× stock solution): Add about 950 mL of water to a 1 L beaker and place on a stir plate with a stir bar. Slowly add 60.6 g of Trizma® hydrochloride, 13.9 g of Trizma® base, and 87.7 g of NaCl and allow to dissolve into solution. Adjust pH to 7.4 with 1 M HCl and add additional water to a final volume of 1 L. Store at room temperature (RT) and dilute 1:10 with water to make a 1× working solution.
18. Endogenous peroxidase quenching solution (hydrogen peroxide/methanol): Add 25 mL methanol (MeOH), 25 mL of 30 % hydrogen peroxide (H_2O_2), and 200 mL TBS, prepared fresh, and discard unused portion.
19. Blocking buffer: Mix 900 μL TBS and 100 μL normal donkey serum, and then add 20 mg nonfat dry milk and 20 mg bovine serum albumin. Mix gently to prevent protein denaturation but thoroughly to ensure full dispersal and uniformity.
20. Humidity chambers can be purchased through commercial vendors (i.e., Fisher Scientific) or self-assembled (Fig. 1).

3 Methods

Carry out all procedures at RT unless otherwise specified. Perform dehydration, rehydration, and coverslip mounting steps in a properly ventilated fume hood. Label slides with a pencil or a solvent resistant marker. Once the tissue specimen has been rehydrated, ensure that it does not dry out between washes or incubations to avoid high background. Include additional, single-stained tissue slides as controls for each individual chromogen and appropriate negative controls to determine nonspecific binding.

3.1 Immunohistochemistry-MAP2 Staining

1. Fill seven coplin jars with 250 mL of the following solvents or solutions: two jars with xylene, four jars with ethanol of decreasing concentrations (100, 95, 90, 70 %), and the final jar with TBS. Perform a series of sequential incubations as indicated. Take care to minimize carryover of solution from one coplin jar to the next (*see* **Note 5**):
   (a) Xylene—30 min.
   (b) Xylene—5 min.
   (c) 100 % EtOH—5 min.
   (d) 95 % EtOH—5 min.
   (e) 90 % EtOH—5 min.
   (f) 70 % EtOH—5 min.
   (g) TBS—5 min.
2. Prepare endogenous peroxidase quenching solution (H_2O_2/MeOH) shortly before use and incubate slides for 30 min followed by a 5-min incubation in TBS (*see* **Note 6**).
3. Prepare the heat-induced antigen retrieval solution by adding 25 mL of target retrieval solution and 225 mL of H_2O to a coplin jar. Submerge the slide rack into the coplin jar and cover with lid. Carefully place the coplin jar into a hot water bath at 95 °C, ensuring that no water is able to enter the jar. Incubate in hot water bath for 1 h, then remove jar, and let it stand at RT for an additional 30 min. Next, perform three 5-min washes in TBS.
4. Apply a hydrophobic barrier around the tissue specimen on each slide using the PAP pen (*see* **Note** 7).
5. Remove slides from the slide rack and place in the humidity chamber. Add avidin blocking solution dropwise until the entire specimen surface is covered. Incubate for 20–30 min. Briefly rinse samples in TBS and add biotin blocking solution dropwise until the entire specimen is covered. Incubate for 20–30 min.
6. Remove biotin blocking solution by gently tapping the slide in a plastic surface and add blocking buffer without a wash between incubations. Ensure that the volume of blocking buffer is sufficient to cover the entire surface of the tissue specimen (approximately 250 μL). Incubate for 1 h at RT in the humidity chamber.
7. Remove blocking buffer by gently tapping and add primary antibody (anti-MAP2, Chemicon), diluted 1:250, in blocking buffer without a wash between incubations.

8. Ensure that the humidity chamber has adequate moisture by adding additional water if necessary, then cover with lid, and seal with Parafilm. Incubate overnight at 4 °C.
9. Place slides in the rack and submerge in TBS for 5 min. Repeat twice. During washes, dilute biotinylated secondary antibody in TBS at 1:250.
10. Apply diluted secondary antibody and incubate for 1–2 h at RT in the humidity chamber.
11. Prior to the end of the secondary antibody incubation, start preparing the avidin–biotin complex (ABC) according to the manufacturer's instructions (*see* **Note 8**). Let the ABC reagent sit at RT for 20–30 min prior to use.
12. After the secondary incubation, place slides back in the rack and submerge in TBS for 5 min. Repeat two more times. Return slides to the humidity chamber and add approximately 250 μL of ABC to each slide, being careful to cover the entire specimen surface. Incubate for 20–30 min at RT in the humidity chamber. Place slides into rack and wash in TBS for 5 min at RT. Repeat two more times.
13. Prepare the NovaRED™ chromogen immediately prior to use according to the manufacturer's instructions (*see* **Note 9**). Place slides in the humidity chamber and apply 250 μL of NovaRED™ to each sample. It is important to work quickly and accurately time the chromogen development. Depending on the antigen of interest, development times will vary. In our hands, a 3–5-min duration with NovaRED™ results in antigen detection in all samples without over-staining.
14. Return slides to rack and submerge in water for 3 min. Repeat once. After the first stain is completed, the slides are ready to undergo the blocking steps needed before adding the second primary antibody. It is important not to dehydrate the slides or wash with ethanol.

3.2 Immunohistochemistry-CXCR4 Staining

1. Remove slides from the slide rack and place in the humidity chamber. Add avidin blocking solution dropwise until the entire specimen surface is covered. Incubate for 20–30 min. Briefly rinse samples in TBS and add biotin blocking solution dropwise until the entire specimen is covered. Incubate for 20–30 min.
2. Remove biotin blocking solution by gently tapping the slide in a plastic surface and add blocking buffer without a wash between incubations. Ensure that the volume of blocking buffer is sufficient to cover the entire tissue specimen surface (approximately 250 μL). Incubate for 1 h at RT in the humidity chamber.

3. Remove blocking buffer by gently tapping and add the second primary antibody (anti-CXCR4), diluted in blocking buffer without a wash between incubations.
4. Ensure that the humidity chamber has adequate moisture by adding additional water if necessary, then cover with lid, and seal with Parafilm. Incubate overnight at 4 °C.
5. Place slides in the rack and submerge in TBS for 5 min. Repeat twice. During washes, dilute biotinylated secondary antibody (donkey anti-mouse) in TBS at 1:250.
6. Apply diluted secondary antibody and incubate for 1–2 h at RT in the humidity chamber.
7. Prior to the end of the secondary antibody incubation, start preparing the ABC (ABC-AP) according to the manufacturer's instructions. Incubate the slides with the ABC-AP reagent at RT for 20–30 min in the dark.
8. Prepare the VectorBlue chromogen immediately prior to use according to the manufacturer's instructions, ensuring that the chromogen is protected from light (*see* **Note 10**). Place slides in the humidity chamber and apply 250 μL of VectorBlue to each sample. It is important to work quickly and accurately time the chromogen development. Depending on the antigen of interest, development times will vary. In our hands, a 10-min development period results in antigen detection in all samples without over-staining.
9. Return slides to the rack and submerge in water for 3 min. Repeat once. The slides can then be dehydrated and cleansed through a series of ethanol and xylene incubations as indicated below (*see* **Note 11**):
   (a) 70 % EtOH—5 min.
   (b) 80 % EtOH—5 min.
   (c) 90 % EtOH—5 min.
   (d) 100 % EtOH—10 min.
   (e) 100 % EtOH—20 min.
   (f) Xylene—5 min.
   (g) Xylene—5 min
10. Remove slides from the rack, wipe residual xylene from the bottom surface with a kimwipe, tilt slide to allow excess xylene on the top of the slide to bead at a corner, and absorb it with a kimwipe™ (*see* **Note 12**).
11. Apply a drop of an aqueous mounting medium such as VectaMount™ to the center of the tissue specimen. Gently lower the coverslip onto the surface to prevent air bubbles from forming under the glass. Allow slides to air-dry overnight before the mounting medium dries (*see* **Note 13**).

3.3 Multispectral Imaging

The theory behind multispectral imaging, the basic use of the Nuance system and a comprehensive description of the Nuance 2.1 software, can be found at the manufacturer's Web site (note: CRi is now part of PerkinElmer). We will provide instructions for this specific protocol and advise users to accordingly modify the protocol as required for their specific purpose. Briefly, each antibody–chromogen complex has a specific color (i.e., blue or red) and that color has a characteristic absorption spectrum. When two chromogens are used simultaneously in a single stain (with proper blocking) the discrete absorption spectra of each antibody–chromogen complex can be digitally separated or unmixed from one another. This allows one to accurately measure and quantify the degree of co-localization as well as determine the relative amount of each respective protein.

Below, we describe the particular camera settings and steps that are used to provide a relative measure of CXCR4 expression within cortical neurons detected using this IHC protocol.

1. Initial camera settings for Nuance camera are the following:

   Acquire tab: Brightfield.

   Binning: 1 × 1; region of interest (ROI): full.

   Use custom wavelengths and exposers: Activated.

   Filter/wavelength selection: Start: 420, Step: 20, End: 720; Narrow: Activated.

   Convert to optical density: Activated.

2. Using the 40× objective, find a representative area of tissue on the dual-stained slide and take a picture using the "Autoexpose Cube" function. Once completed, position objective over an area of the slide with no tissue and use the "Acquire Reference" function.
3. Next, place the single-stained MAP2 slide on the microscope, focus on the area of interest, and select "Acquire Cube." Under the "Spectra" tab, sample the absorption spectra and label it "MAP2/NovaRed."
4. Repeat the process with the single-stained CXCR4 slide and label the absorption spectra "CXCR4/VectorBlue." Save the spectral library as "NovaRed/VectorBlue Library," and protocol settings (*see* **Note 14**).
5. Once completed, the dual-stained slides can be imaged using the "Acquire Cube" function and unmixed using the spectral library (Fig. 2).
6. Once the image cube is unmixed, four images will be present: the unmixed image cube (Fig. 3a), a grayscale image of the unmixed NovaRed MAP2 spectrum (Fig. 3b), a grayscale image of the unmixed VectorBlue CXCR4 spectrum (Fig. 3d), and a composite, pseudo-color image of the two unmixed spectra (Fig. 3f).

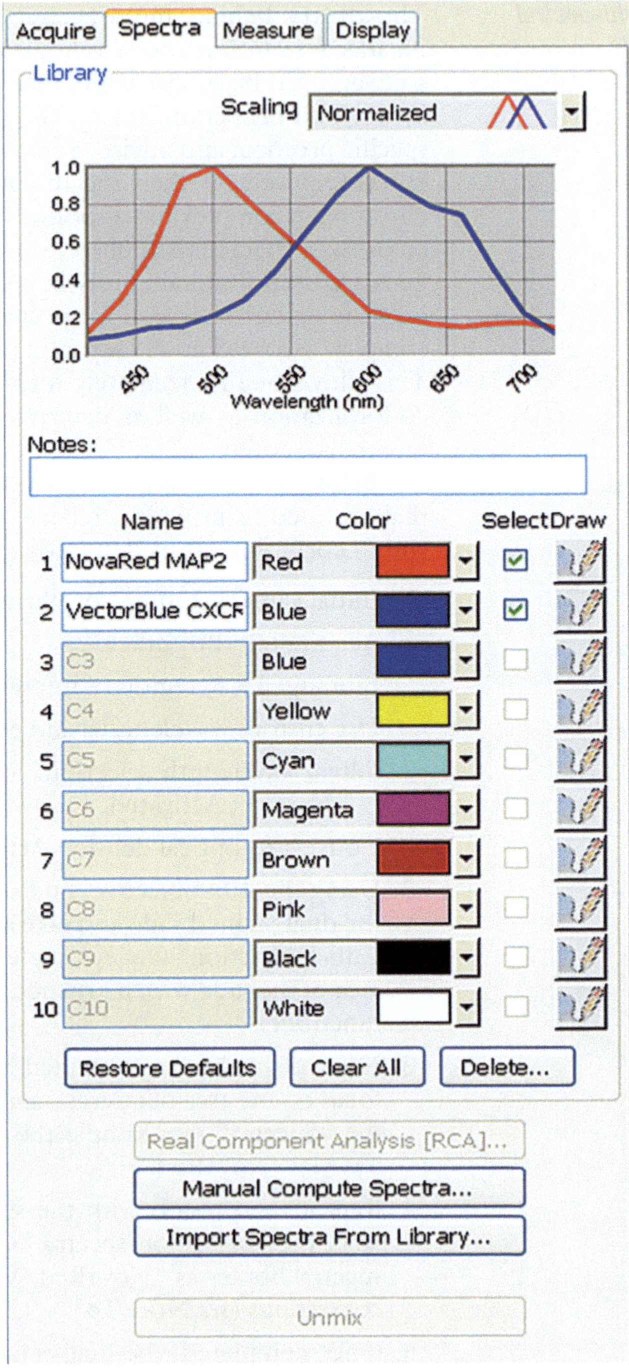

Fig. 2 Screen shot of Nuance software showing the spectral library of each chromogen. Chromogens, unlike fluorophores, have broad absorption spectrums. In order to maximize distance between peak absorbances, we chose NovaRed and VectorBlue represented by the *red* and *blue curve*, respectively. Properly creating the spectral library is the most important factor in ensuring good, consistent results

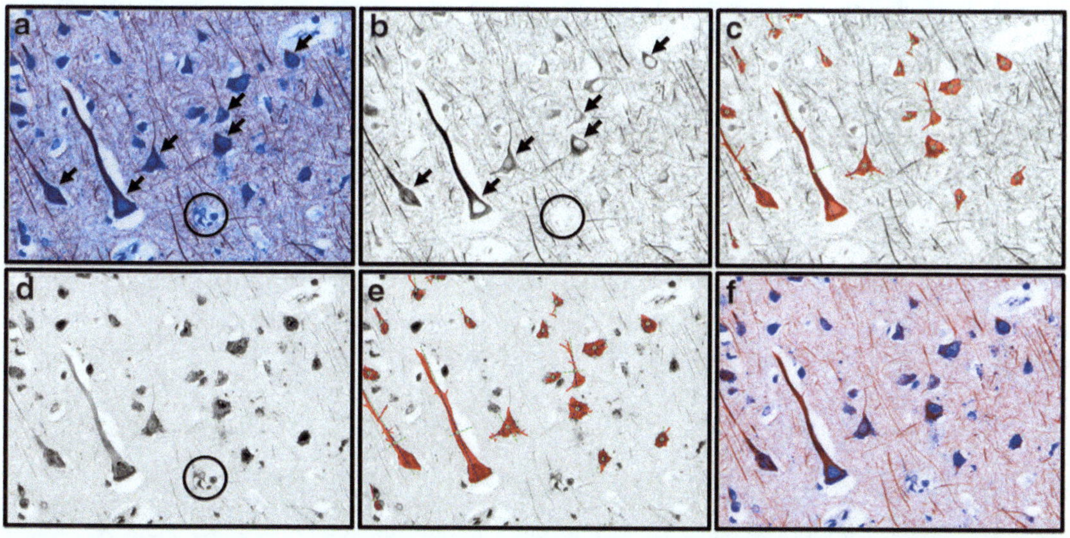

Fig. 3 Multispectral image analysis of postmortem human cortex. Archived sections of human cortex (normal brain) were dual stained with antibodies for a neuronal marker, MAP2 in *red*, and the chemokine receptor, CXCR4, in *blue*. (**a**) A standard color photomicrograph, taken at 40× magnification, shows expression of CXCR4 within neurons (a subset of which are indicated by *arrows*) and, to a lesser extent, in neighboring glia (a subset of which is indicated by the *black circle*). (**b**) Spectral unmixing shows absorption from only the MAP2:NovaRed chromogen complex. The neurons are identified by the high optical density relative to background (a subset of which are indicated by *arrows*). The indicated glia have no expression of MAP2 (indicated by the *black circle*). (**c**) Neurons are digitally isolated based on the optical density of the MAP2:NovaRed chromogen complex. They are indicated by *red* cropping and numbered. (**d**) Spectral unmixing of the CXCR4:VectorBlue chromogen complex shows a mixed population of cells including neurons and glia (a subset of which are indicated by the *black circle*). (**e**) The area marking MAP2+ neurons is overlaid by the CXCR4:VectorBlue optical density. The optical density values reported in the analysis reference CXCR4:VectorBlue optical density within the MAP2:NovaRed cropped regions. (**f**) A pseudo-colored composite image with MAP2+ (*red*) and CXCR4 (*blue*)

7. Select the NovaRed MAP2 spectrum window (Fig. 3a), and open the measure tab of the Nuance software. Set the minimum connected pixel value to 1,000. Using the OD for MAP2, set the threshold (automatically or manually) to identify the ROI (i.e., cortical neurons). The result will be a digital cropping of the MAP2+ cortical neurons (Fig. 3c) (*see* **Note 15**).
8. Once the neurons have been digitally cropped as ROI, save measurement regions (using the right mouse click pull-down command list). Open the VectorBlue CXCR4 spectrum window (Fig. 3d) and paste the saved measurement regions.
9. Now, the ROI, which correspond to the neurons, are superimposed over the VectorBlue CXCR4 spectrum (Fig. 3e). Each ROI is now represented with a series of region measurements (Table 1). A portion of the region measurements is presented below (adapted from CRi user's manual for Nuance 2.1.).

Table 1
Optical density data from regions of interest (ROI) identified in Fig. 3

ROI#	Total signal (OD)	Average signal (OD)	Standard deviation (OD)	Max signal (OD)	Area (pixels)	Area (μm^2)	Major axis	Minor axis	*X* location	*Y* location	Spectrum ID	Threshold	Min connected
1	4552.29	0.268873	0.0935855	0.562295	16,931	65093.91	598.726	96.724	383.43	644.019	VectorBlue CXCR4 40×	0	1,000
2	2218.21	0.237623	0.111994	0.53121	9,335	35889.89	399.819	81.406	131.259	687.221	VectorBlue CXCR4 40×	0	1,000
3	1487.03	0.282597	0.100695	0.488881	5,262	20230.59	141.88	137.55	583.211	621.952	VectorBlue CXCR4 40×	0	1,000
4	1580.72	0.318116	0.1125	0.58583	4,969	19104.11	105.064	81.3826	864.734	545.146	VectorBlue CXCR4 40×	0	1,000
5	1543.66	0.35131	0.0771617	0.536753	4,394	16893.43	107.716	65.3602	932.881	296.536	VectorBlue CXCR4 40×	0	1,000
6	636.222	0.189296	0.101757	0.444176	3,361	12921.9	214.286	84.966	852.98	390.393	VectorBlue CXCR4 40×	0	1,000
7	1184.04	0.378651	0.114355	0.603569	3,127	12022.25	83.2953	54.8585	1149.97	197.181	VectorBlue CXCR4 40×	0	1,000
8	1030.03	0.423358	0.118829	0.605729	2,433	9354.054	84.3596	56.4143	1208.73	667.386	VectorBlue CXCR4 40×	0	1,000
9	796.435	0.329242	0.0712924	0.489915	2,419	9300.229	76.7109	50.346	27.5308	405.687	VectorBlue CXCR4 40×	0	1,000
10	973.827	0.408827	0.11675	0.612854	2,382	9157.977	89.9736	52.6041	969.739	749.785	VectorBlue CXCR4 40×	0	1,000
11	514.622	0.252885	0.103562	0.50358	2,035	7823.88	117.308	36.9643	74.6747	195.713	VectorBlue CXCR4 40×	0	1,000
12	698.828	0.364163	0.0738177	0.551445	1,919	7377.899	72.7532	45.882	724.166	227.858	VectorBlue CXCR4 40×	0	1,000
13	506.677	0.296997	0.0987565	0.483302	1,706	6558.987	99.7813	32.7905	422.746	187.652	VectorBlue CXCR4 40×	0	1,000
14	438.033	0.26261	0.151042	0.521978	1,668	6412.89	109.748	63.541	214.25	63.503	VectorBlue CXCR4 40×	0	1,000
15	362.871	0.238574	0.0991011	0.424898	1,521	5847.726	103.633	52.8815	778.871	110.283	VectorBlue CXCR4 40×	0	1,000
16	452.315	0.307071	0.0752889	0.452115	1,473	5663.182	65.0203	42.9909	766.887	416.155	VectorBlue CXCR4 40×	0	1,000
17	361.788	0.293183	0.123844	0.485235	1,234	4744.309	113.935	24.1783	51.4733	919.497	VectorBlue CXCR4 40×	0	1,000
18	347.232	0.31114	0.128388	0.475731	1,116	4290.639	82.6132	37.5555	294.694	9.87007	VectorBlue CXCR4 40×	0	1,000

Each individual cropped neuron is defined as a region of interest, or ROI, and numbered

Region Measurements:

ROI. The numerical label assigned to each object (1, 2, 3, etc.).

Total Signal (OD). The sum of OD values for each pixel within an ROI.

Average Signal (OD). The average OD per pixel within an ROI.

Standard Deviation (OD). A measure of variation in OD within an ROI.

Max Signal (OD). The greatest OD value within a single ROI.

Area (pixels). The number of pixels within an ROI.

Area (μm^2). The number of square microns within an ROI. This requires additional calibration.

Major Axis. The length of the minimum area bounding box enclosing the region.

Minor Axis. The width of the minimum area bounding box enclosing the region.

X Location. Is the center of gravity's *x*-coordinate.

Y Location. Is the center of gravity's *y*-coordinate.

Spectrum ID. The spectrum (from the spectral library) used to unmix the image.

Threshold. Pixels with an OD value below the set threshold will be ignored. We generated our measurements using a threshold value of zero and compared the fold-difference between OD values between two conditions.

Min. connected. A size threshold set by the user on the smallest number of continuous pixels allowed into the analysis. No ROI will have an area (in pixels) below the minimum connected threshold. This is a useful way to eliminate small artifacts within the visual field.

10. We use *Average Signal (OD)* of CXCR4 (or other protein of interest) as the primary measurement endpoint, but other measures such as *Max. Signal* or *Total Signal* may be more appropriate depending on the specific question being addressed.
11. Analyze staining. For example a pseudo-colored composite of the dual staining demonstrates strong, punctate neuronal CXCR4 staining in the cell body (Fig. 4).

3.4 Laser Capture Microdissection

LCM has become an important technique for harvesting specific tissue regions or cell populations for protein or RNA isolation. When applying this technology to large-scale studies, automation becomes a necessary component. We have developed a strategy for automated neuron identification and capture and present guidelines for adapting our technique to fit broader applications. Importantly, this protocol assumes that the user has a working knowledge of LCM, software interface, and routine functions such as laser and

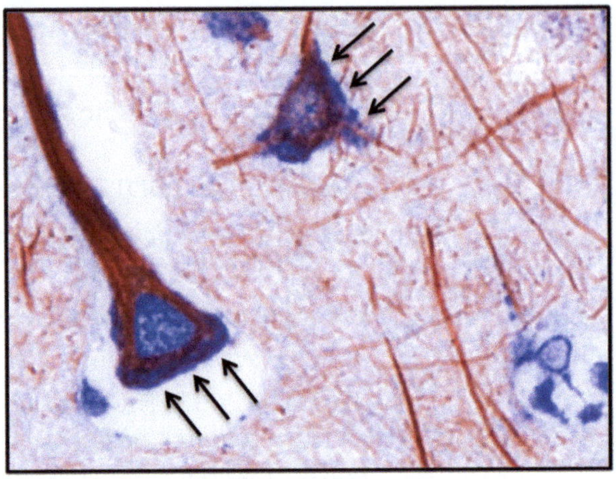

Fig. 4 Composite image of human cortical neurons dual stained for MAP2 and CXCR4. Representative, pseudo-colored image of human cortex (normal brain) dual stained for the neuronal marker MAP2, *in red*, and the chemokine receptor CXCR4, *in blue*. The image demonstrates MAP2 staining of neuronal cytoplasm with punctate expression of CXCR4, predominantly along the cytoplasmic membrane as indicated by the black arrows

stage calibration. Our purpose is to detail the process of automating object identification and capture, using the common neuronal marker, MAP2.

1. Slides are prepared as described in Subheading 3.1. Following development with NovaRed, rinse slides thoroughly in RNAse-free water.
2. Open PALM RoboSoftware, select appropriate camera (e.g., AxioCamICc1), and set light intensity to *3,200 K* (*see* **Note 16**). Ensure that the light aperture is centered and fully opened before setting the *White Balance* and *Shading Control*. Light intensity can be adjusted with polarized filters or *Analog Gain* without changing the relative color composition and should be used in place of *Light Intensity %*.
3. Take a standard red–green–blue (RGB) image of a representative slide preparation in the PALM RoboSoftware suite at the appropriate magnification (*see* **Note 17**). Save this file to the desktop as it will be used to fine-tune the script outlined in Fig. 5. For this protocol, we will name this image "Novared MAP2+ 20×" to indicate the chromogen, marker of interest, and magnification.
4. Open the AxioVision software and enter the *Script Editor*. Enter the commands, thresholds, and parameters outlined in Fig. 5. The script is applied to a single neuron in Fig. 6 or a collection of neurons in Fig. 7. The ability to eliminate extraneous regions, based on color intensity or region size, is essential to prevent glial contamination. The script can be tested and

1. Command: Sigma

Parameters: Kernel Size=3

Sigma=27

2. Command: Threshold interactive (HLS), Create Binary Image

Parameters: Level Low Hue=0

Level High Hue=28

Parameters: Level Low Luminescence=0

Level High Luminescence=178

Parameters: Level Low Saturation=24

Level High Saturation=255

3. Command: Clean

Parameters: Min=150

Max=10,000

4. Command: Dilate

Parameters: Octagon, Count 8

5. Command: Measure Binary Image, Element type Freehand

6. Command: Start Measurement

Fig. 5 AxioVision-generated script for identifying MAP2+ cortical neurons labeled with NovaRed chromogen. A script can be generated using the five commands, with accompanied parameters. While the commands are essential, the parameters may be modified by the end user based on staining intensity, magnification, and any artifacts of fixation that might exist

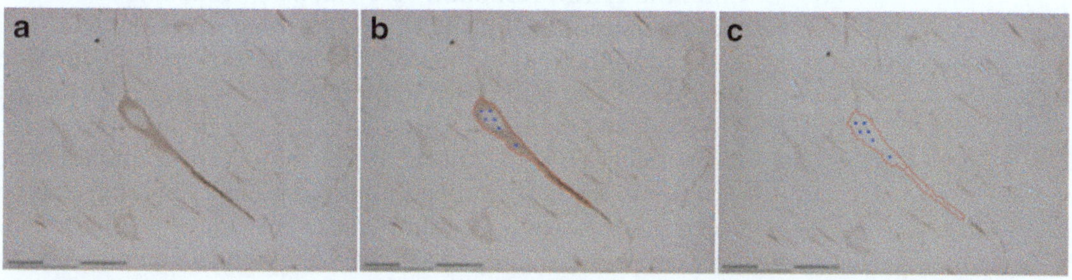

Fig. 6 Script-based laser capture microdissection of cortical neuron (40×): (**a**) Human cortex stained with MAP2 conjugated with NovaRed as described in Subheading 3.1. (**b**) Application of the cortical neuron capture script outlined in Fig. 5 was applied and identified in the neuronal boundary to be cut (*red boundary*) and points for pressure catapult (*blue dots*). (**c**) The post-capture image shows a complete removal of the script-defined region

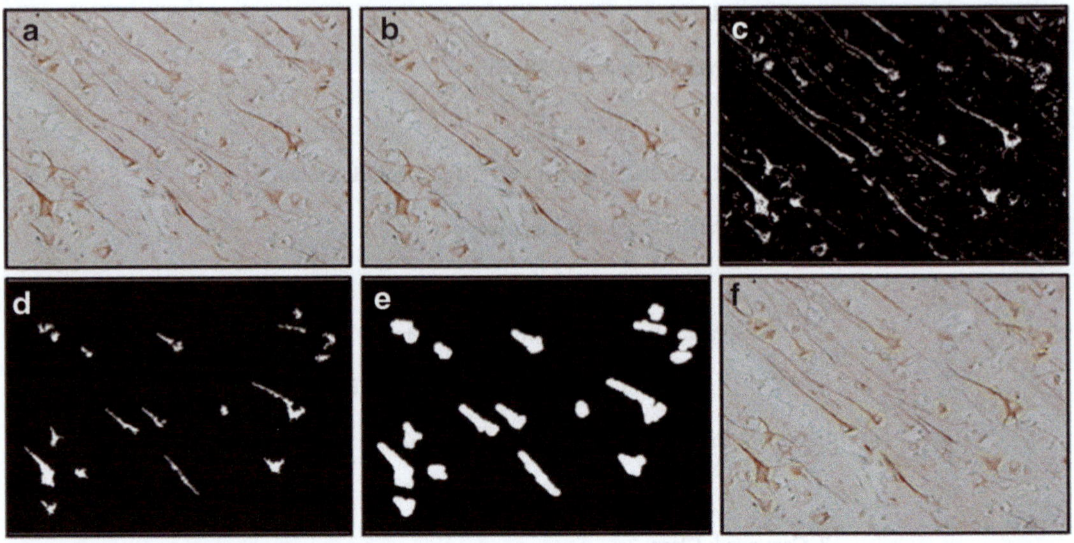

Fig. 7 Script-based laser microdissection at 20×: (**a**) RGB image of human cortex stained with MAP2. (**b**) Image following Sigma command. (**c**) Binary selection-based Hue/Luminosity/Intensity thresholds calibrated to 3,200 K. (**d**) Size gating between 150 and 10,000 continuous pixels. (**e**) Pixel dilation. (**f**) Image capture boundary (*in yellow*) superimposed over original RGB image

modified by the user in the software program, depending on slight variations in the protocol or specific interests of the user. Indicate the chromogen, marker of interest, and magnification to prevent confusion with other scripts. Save the script as "Capture NovaRed MAP2+ 20×.ziscrpit." Close AxioVision.

5. Open PALM RoboSoftware and import "Capture NovaRed MAP2+ 20×.ziscrpit" into the *Field of View (FoV) Analysis* pull-down menu by accessing *Adjustments* menu -> *PalmRobo...* -> *FoV Script Assignment*. The FoV Script Assignment allows you to assign the custom script "Capture NovaRed MAP2+ 20×.ziscrpit."

6. Enter the PALM RoboSoftware *Navigator* function and define the rectangular ROI using the *Set ROI-top-left* & *Set ROI bottom-right* functions. Under *AxioVision Analyze* tab, use the *Analyze scripts* pull-down tab to select "Capture NovaRed MAP2+ 20×." Select *Analyze ROI* to begin the scan. Depending on the area of the ROI, magnification, and number of objects, the scan can last multiple hours. This step can be scaled up or down depending on the amount of RNA needed. An example of a completed scan with pseudo-colored objects (i.e., MAP2+ neurons) is seen in Fig. 8.

7. Optional RNA amplification: The amount of RNA collected can vary greatly depending on the condition of the material used. In cases where the yield is insufficient, we have used a commercially available mRNA amplification kit (e.g., SMART mRNA Amplification Kit, Clontech).

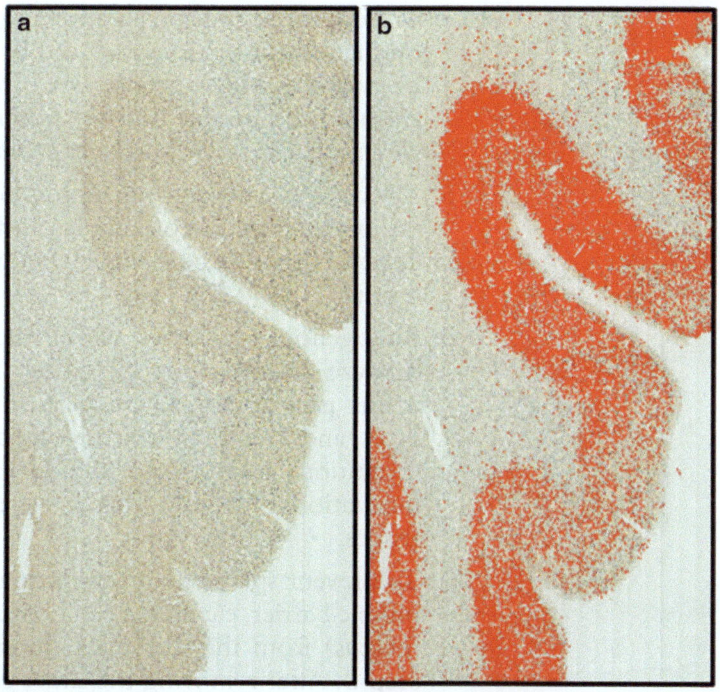

Fig. 8 Automatic neuron identification in human cortex. (**a**) Over 1,500 photomicrographs (at 20× magnification) are systematically scanned and analyzed for the presence of MAP2+ cortical neurons using the script demonstrated and detailed in Subheading 3.4. (**b**) MAP2+ neurons are pseudo-colored in *red*. The automated detection and capture typically result in over 10,000 neurons, collectively covering an area of 6–8 mm^2 and yield up to 20 mg of RNA, following amplification, for gene expression analyses and archiving

4 Notes

1. Xylene substitutes (i.e., Histosolve) are commonly used petroleum-based solvents that have a reduced risk of inhalation toxicity. For the sake of clarity, we refer to xylene substitutes simply as xylene.
2. Denatured EtOH may contain additives capable of interfering with some staining protocols or downstream applications. If using denatured alcohol, check the specific additive to ensure that it is compatible.
3. The serum used in the blocking buffer should be from the same species used to create the secondary antibody. Therefore, we use normal donkey serum and a biotin-conjugated donkey anti-mouse secondary antibody.
4. Select chromogens to prevent overlap in the absorbance spectrum. The broad absorption spectra can overlap and introduce error during multispectral image (MSI) analysis. Because of this, the number of chromogens that can be used simultaneously

within a single slide preparation is generally limited to three. In our hands, the error introduced by adding a third chromogen negates any added benefit. We strongly recommend limiting chromogens to two in most instances. *Importantly, treat all chromogens as carcinogenic and dispose of accordingly.*

5. Xylene substitute and EtOH solutions can be reused (depending on the number of samples and time between uses); however, we recommend using fresh solutions each time for optimal results.

6. Endogenous peroxidase within the tissue may nonspecifically catalyze chromogen precipitation. The "quenching" of endogenous peroxidase can be performed with an incubation of H_2O_2 and MeOH is added to enhance the reaction. The concentration of H_2O_2 and MeOH can be increased to shorten incubation time, but this may result in damage to fragile tissues.

7. To prevent specimens from drying while applying the hydrophobic barrier, each slide can be individually withdrawn (using forceps) from the slide rack during the washes described in **step 3**. After removing the slide, gently tap it on a plastic surface to quickly remove excess TBS. Apply the PAP pen to the dry glass surface, while the tissue still retains moisture and returns back to the slide rack. Ensure that the hydrophobic barrier is dry before submerging in TBS.

8. The ABC greatly increases the sensitivity of antigen detection; therefore it is recommended but not required depending on the abundance and retrieval of a given antigen. In order to prevent cross-reaction between the two avidin–biotin amplifications steps (ABC and ABC-AP), we use a HRP-based ABC for the first antigen and an alkaline phosphatase-based ABC for the second antigen. HRP-based ABC is more sensitive and results in a sharper chromogen precipitate, while the alkaline phosphatase-based ABC is less sensitive and results in a more diffuse precipitate. We use the HRP-based ABC first, to amplify the more abundant antigen, as some precipitate will be washed away during subsequent washes and incubations. Importantly, some chromogen substrates, most notably 3, 3-diaminobenzidine (DAB), form a shell around the antigen of interest that "shields" it from antibodies. The shielding is a result of the unusually small pore size of the precipitated substrate. For this reason, DAB should not be used first in a dual-stain protocol where co-localization might occur.

9. The water used to make the working solution of NovaRed must be free of oxidizable carbon. This can vary within a laboratory over time and from laboratory to laboratory even when identical water purification systems are used. To prevent

interference with the oxidization and variability between batches, we do not use water from our on-site purification system when making H_2O_2 substrate. Instead, we use a single source of purified, distilled water from a commercial source (i.e., Gibco).

10. The ABC-AP complex is for use with alkaline phosphatase chromogens only. These chromogens require levamisole in the working solution to act as a competitive inhibitor of endogenous phosphatases within the central nervous system. Adding 0.1 % Tween 20 can improve the staining clarity.
11. Using fresh stocks of EtOH and xylene are particularly important during the dehydration and cleansing steps. The appearance of small "bubbles" uniformly present under magnification suggests that the EtOH needs to be changed.
12. Do not use paper towels or other absorbent materials that leave fibers or debris on the slide surface. In our experience, Kimwipes™ work best.
13. VectaMount™ is a permanent mounting medium, but some chromogens require aqueous-based mounting medium such as VectaMount™ AQ. The coverslip of a dried slide can be removed by submerging in xylene overnight without damaging the tissue.
14. These settings are specific for the 40× objective and camera parameters used while generating the library. If software alterations are made, such as wavelength range, then the library will not be loaded. The spectral library can only be used to unmix images of the same magnification. If you want to unmix images at a different magnifications (e.g., 20× and 5×), then the appropriate libraries must be used (e.g., 20× and 5× libraries, respectively).
15. The criteria used to define the ROI should be as consistent as possible. We use the MAP2 spectrum OD and neuronal morphology to ensure that only neurons are selected. Occasionally, blood vessels, connective tissue, or artifacts of fixation can result in extremely high OD values. These cases can be filtered out individually by hand or based on size, by eliminating all ROI with a minimum number of connected pixels.
16. We use a set color temperature of 3,200 K to maintain consistency across image analyses. The script parameters for *Threshold Interactive (HLS)* described in Fig. 5 are calibrated to 3,200 K but can be adjusted independently with the PALM RoboSoftware directly using the *FoV Analysis*.
17. Script parameters for *Sigma*, *Clean*, and *Dilate* are influenced by magnification and must be modified for each different magnification. If you will work at 20× and 40× magnifications, you will need two separate scripts.

Acknowledgements

This work was supported by National Institute of Health—National Institute on Drug Abuse Grants DA15014 and DA19808 (OM). Jonathan Pitcher was a student fellow of the neuroAIDS training grant Interdisciplinary and Translational Research Training Program in neuroAIDS (T32MH078795).

References

1. Cattoretti G, Pileri S, Paravicini C, Becker MH, Poggi S, Bifulco C, Key G, D'Amato L, Sabattin E, Feudale E (1993) Antigen unmasking on formalin-fixed, paraffin-embedded tissue sections. J Pathol 171:83–98
2. Boenisch T (2005) Effect of heat induced antigen retrieval following inconsistent formalin fixation. Appl Immunohistochem Mol Morphol 13:283–286
3. Buchwalow I, Bocker W (2010) Immunohistochemistry: basics and methods. Springer-Verlag, Berlin/Heidelberg

Chapter 4

Chemokine Receptors and Neural Stem Cells

Meizhang Li

Abstract

Neural stem cells (NSCs) represent a limited population of progenitor cells in the central nervous system that sustain their self-renewal and multipotency from early development to adulthood. Recent evidence suggests that chemokine receptors are constitutively expressed by NSCs and are directly involved in stem cell biology. As cell surface receptors, chemokine receptors also provide an important avenue to enrich these cells and further identify the potential molecular pathways required to maintain their biological functions. Here, I describe in vitro methods that have been widely applied to sort, culture, maintain, and differentiate NSCs.

Key words Neural stem cells, Chemokine, CXCL12, CXCR4, Neurospheres, Fluorescence-activated cell sorting, Differentiation, Proliferation

1 Introduction

Chemokines are a group of small secreted proteins that have been classified according to two conserved cysteine residues near the N-terminus. These are divided into four subfamilies: CXC, CC, C, and CX3C subfamilies [1]. Chemokine receptors are conserved seven-transmembrane G-protein-coupled receptors (GPCRs) [2–4]. Binding of chemokines to their receptors is known to facilitate the migration of leukocytes to target tissues or organs by inducing G-protein-mediated downstream signaling pathways [5].

In the central nervous system (CNS), CXCL12–CXCR4 signaling has been implicated as a key regulator of cell migration, axon guidance, as well as maintenance of neural progenitor cells (NPCs) [6]. Pathologically enhanced CXCL12–CXCR4 signaling may also promote the formation of new vessels through recruiting circulating endothelial progenitor cells or directly increase endothelial cell proliferation or migration [7]. In the adult mouse subventricular zone (SVZ) where endogenous neural

Astrid E. Cardona and Eroboghene E. Ubogu (eds.), *Chemokines: Methods and Protocols*, Methods in Molecular Biology, vol. 1013, DOI 10.1007/978-1-62703-426-5_4, © Springer Science+Business Media New York 2013

stem cells (NSCs) are located, a recent study demonstrated that proliferating NPCs can home to peri-vascular niches in a CXCL12–CXCR4 pathway-dependent manner [8].

Several other studies have shown that CXCR4 is expressed in early NPCs and CXCL12–CXCR4 signaling regulates the migration and proliferation of NPCs [9–14]. However, it still remains a challenge to directly isolate NSCs due to the lack of useful cell surface markers. We recently observed the expression of CXCR4 in radial glia cells located in the early developing ventricular zone [15]. We further identified CXCR4-expressing NSCs, demonstrating the self-renewal and multipotent capabilities in vitro. In this chapter, I describe methods for sorting and enriching NSCs using anti-CXCR4 antibody. In addition, I also describe some basic NSC-related experiments such as culture, maintenance, differentiation, and proliferation.

2 Materials

Prepare stock solutions, make necessary aliquots, and store all reagents in −80 °C for long-term use.

1. Full neurobasal (NB) medium (*see* **Note 1**): 10 mL B27 (Gibco-BRL), 10 μg epidermal growth factor (EGF) (Invitrogen), 10 μg basic fibroblast growth factor (FGF) (Invitrogen), 5 mL 1× sodium pyruvate (Gibco-BRL), 2 mM L-glutamine (Gibco-BRL), 50 units/mL Penicillin–Streptomycin (Gibco-BRL).
2. Differentiation culture medium: DMEM/F12 medium supplied with 1 % (v/v) fetal bovine serum (FBS), N2 supplement (Gibco-BRL), and 10 ng/mL platelet-derived growth factor (PDGF) (R&D) to induce oligodendrocyte differentiation. To differentiate NSCs into astrocytes, NSCs are added into DMEM/F12 medium with 10 % (v/v) FBS. To differentiate NSCs into neuron-lineage cells, NSCs are cultured into null NB medium supplied with B-27 (Gibco-BRL), 1 % penicillin–streptomycin, and 1 % L-glutamine (Gibco-BRL).
3. Freezing medium: 80 mL 10 % heat-inactivated fetal bovine serum (Gibco-BRL), 10 mL dimethyl sulfoxide (DMSO) (Fisher), 10 mL Null NB medium (Gibco-BRL).
4. FACS buffer: Hank's balanced salt solution (HBSS), without phenol red (Gibco-BRL), 1 % bovine serum albumin (BSA) (Gibco-BRL), 0.1 % sodium azide (Fisher), 1 mM EDTA (Fisher). 7.5 % BSA stock is prepared with 1× phosphate-buffered saline (PBS) and diluted further with 1× PBS to obtain a final concentration of 1 %.

5. Poly-D-lysine (PDL): Take 0.5 mL 100× PDL (Sigma) stock, dissolve into 49.5 mL distilled water (final concentration is 1× PDL), filter sterilize the solution through a 0.22 μm unit, and store at 4 °C.
6. CXCL12 stock: Reconstitute lyophilized mouse recombinant CXCL12 (R&D) in sterile 1× PBS containing 0.1 % BSA to 10 μg/mL.
7. Other reagents: Mouse monoclonal anti-human CXCR4 antibody 12G5 (1:200 dilution) and isotype (1:200 dilution) are prepared with FACS buffer; 1 mg/mL propidium iodide (PI) is prepared with distilled water and diluted 1:1,000 with FACS buffer to generate its working solution; 12.5 mg/mL Hoechst 33342 stock is prepared with water and diluted 1:1,000 with FACS buffer to generate its working solution; Pyronin Y (Sigma).
8. CyQUANT® NF assay kit (Invitrogen).

3 Methods

3.1 Culture NSCs

1. Euthanize adult mice with CO_2 and sacrifice the animals.
2. Dissect out the brains [4–6] in a 10 cm diameter Petri dish.
3. Wash the brains three times with 5–10 mL HBSS medium, and then aspirate the medium.
4. Section the brains to obtain 500–700 μm slices, isolate the tissue from SVZ, and mince into a slurry with scissors at room temperature (RT).
5. Treat the slurry with 1 mL 0.05 % trypsin–EDTA at 37 °C for 30 min at RT.
6. Terminate the slurry digestion with 1 mL DMEM medium containing 10 % heat-inactivated FBS.
7. Incubate the mixture at 37 °C for 30 min (*see* **Note 2**).
8. Prepare a 50 mL conical tube with a 70 μm nylon cell strainer placed atop.
9. Gently triturate cells ten times (*see* **Note 3**) and pass the mixture through the cell strainer (*see* **Note 4**).
10. Use 1–2 mL null NB medium to wash the cell strainer to collect adherent cells.
11. Centrifuge pooled suspension at 1,000 × *g* for 5 min at RT to pellet the cells.
12. Resuspend the cells with 10 mL 1× PBS and immediately mix with 30 mL water.

13. Quickly spin down cells (1,200 × *g* for 3 min at RT) and remove the supernatant.
14. Determine cell count using a hemacytometer.
15. Resuspend 1 million cells in 10 mL full NB medium and culture the cells at 37 °C in a new 10 cm diameter Petri dish (*see* **Note 5**).

3.2 Passaging NSCs

1. Neurospheres are not adherent, so they can be easily collected to be centrifuged at 1,000 × *g* for 5 min at RT.
2. Aspirate the supernatant, leaving the neurosphere pellet behind.
3. Add 1 mL 0.05 % trypsin and gently resuspend the pellet.
4. Incubate neurospheres for 3–5 min at 37 °C (*see* **Note 6**).
5. Gently triturate the cell suspension and add an equal volume of DMEM medium containing 10 % FBS to neutralize the trypsin.
6. Add 10× volume of null NB medium to dilute the serum.
7. Centrifuge the cell suspension at 1,000 × *g* for 5 min at RT.
8. Remove supernatant and resuspend cells in at least 10 mL null NB medium to wash out the serum.
9. Repeat centrifugation (*see* **step** 7) and aspirate the supernatant.
10. Resuspend the cell pellet in 10 mL full NB medium for further culture (*see* **Note** 7).

3.3 NSC FACS Using CXCR4

1. Pass cell suspension through a 35 μm nylon cell strainer to break down large cell clusters.
2. Centrifuge at 1,000 × *g* for 5 min at RT to collect the cells.
3. Remove supernatant and wash the cells with 10 mL of null NB. Quantify cells (Subheading 3.1, **step 14**). Repeat centrifugation (**step 2**), and aspirate supernatant.
4. Resuspend 10 million cells with 500 μL 5 % BSA in FACS buffer.
5. Stain the cells with CXCR4 antibody (1:200 dilution in FACS butter) or isotype control (1:200 dilution in FACS buffer) for 30 min at RT on a shaker (*see* **Note 8**).
6. Centrifuge cells (centrifuge cells at 1,000 × *g* for 5 min at RT) and remove the supernatant.
7. Wash the cells twice with FACS buffer and resuspend in 200–500 μL FACS buffer (*see* **Note 9**).
8. Keep cells on ice and protect from light.
9. Add 1 μL of PI dye into the cell suspension prior to FACS (*see* **Note 10**).

3.4 Differentiation of NSCs In Vitro

1. Coat culture dishes or plates with 5.0 μg/cm² PDL solution.
2. Make sure that enough PDL solution covers the entire surface of the dishes or the plates.
3. Incubate culture dishes or plates at RT in a sterile biosafety cabinet.
4. Aspirate residual PDL solution and rinse the surface thoroughly with sterile water.
5. Dry the dishes or the plates for 1–2 h (*see* **Note 11**).
6. Dissociate neurospheres by following the procedures described as above (Subheading 3.2).
7. Wash the cells once with null NB medium (*see* Subheading 3.2, **steps 6** and 7), resuspend the cells in 500 μL null NB medium, and enumerate cells (*see* Subheading 3.1, **step 14**).
8. Incubate the cells with differentiation medium (*see* Subheading 2, **item 2**) on PDL-coated dishes or plates at a density of $1–2 \times 10^5$ cells per plate.
9. Check cell growth and differentiation every day; neurospheres will collapse after NSC differentiation, differentiated cells will look adherent and morphology characteristic of oligodendroglia.

3.5 NSC Proliferation Analysis

1. Dissociate cells from neurospheres (*see* Subheading 3.2). Reconstitute in DMEM/F12 medium supplemented with 10 % FBS, aiming for a final cell density of 1.25×10^4 to 1.0×10^5 cells/mL.
2. Seed 100 μL of the cell suspension per well on PDL-coated 96-well plates.
3. Add 1–20 ng/mL recombinant mouse CXCL12 to the culture in DMEM/F12 medium.
4. Perform CyQUANT® NF assay (Invitrogen) to analyze the cell proliferation at different time periods according to the manufacturer's instructions.
5. Stain the cells with DNA dye Hoechst 33342 for 30 min–1 h at 37 °C to allow good penetration within the neurosphere.
6. Measure fluorescent intensity by using a fluorescence microplate reader (*see* **Note 12**).

3.6 NSC Cell Cycle Analysis

1. Collect and dissociate neurospheres by following the procedure described previously (*see* Subheading 3.2).
2. Wash the cells once with null NB medium and remove the supernatant (*see* Subheading 3.2, **steps 6** and 7).
3. Resuspend 50,000–100,000 cells in 1 mL null NB medium.
4. Add 12.5 μg/mL Hoechst 33342 (Invitrogen) to the cell suspension for 45 min at 37 °C.

5. Wash the cells once with null NB medium and remove the supernatant (*see* **step 2**).
6. Resuspend the cells once again in 1 mL null NB medium.
7. Stain the cells with 0.5 µg/mL Pyronin Y (Sigma) for 45 min at 37 °C (*see* **Note 13**).
8. Wash the cells twice with null NB medium and resuspend in 1 mL as previously described (*see* **steps 3** and **6**).
9. Analyze cell cycle characteristics using a flow cytometer (e.g., BD Biosciences LSRII) equipped with 325 nm ultraviolet and 532 nm green lasers (*see* **Note 13**).

4 Notes

1. To prepare full NB medium, add all reagents in 500 mL neurobasal medium, referred to as null NB medium.
2. Incubate for longer periods if the tissue is very large.
3. Triturate cells again, being careful not to generate bubbles, and deliver into a new conical tube with a 70 µm nylon cell strainer atop.
4. If needed, incubate longer until slurry is a homogenous liquid when triturated. Pipette tips are used to aid in filtration by dissociating bigger clumps of cells (help cells through filter).
5. Small-sized neurospheres (aggregates) will be observed after 2–3 days.
6. Add 1 mL for each plate harvested directly to the cell pellet. Passage cells at least once every five days when neurospheres are forming in culture; otherwise they get too large and start to differentiate. Neurospheres are considered too large if they appear dark in the center.
7. If FACS will be performed after this, dissociated cells should be allowed to recover in full NB medium for 3–4 h at 37 °C prior to sorting.
8. Prepare three round-bottom FACS tubes: one for antibody labeling (i.e., CXCR4), one for the isotype control, and one for an unstained (background) control. Split 50 µL of the cell suspension into each of the control tubes (isotype and unstained). Use the rest for labeling with the antibody of interest.
9. Vary the volumes appropriately according to the cell count.
10. For tube sorting, label two 5 mL FACS cluster tubes or 15 mL conical tubes per sample and add medium (3 mL for 5 mL cluster tubes, 5 mL for 15 mL conical tubes). Cells should be

sorted into full NB medium. For 96-well plate sorting, add 200 μL full NB medium into each well prior to sorting.

11. Alternatively, store the dish or the plate at 4–8 °C or RT for up to three months.
12. For example, SpectraMax M2 from Molecular Devices, with the following parameters: excitation at ~485 nm and emission detection at ~530 nm. Fluorescence measurements are indicative of cell numbers according to the linear range of the standard curve.
13. Hoechst 33342 DNA dye is used to outline the major population distributions of NSCs in the G1/G0 phases and G2/S/M phases. Cells in G1/G0 phases are further distinguished based on differential uptake of Pyronin Y RNA dye. Quiescent cells, which are arrested in G0 phase, have lower RNA levels compared to active G1 cells.

References

1. Zlotnik A, Yoshie O (2000) Chemokines: a new classification system and their role in immunity. Immunity 12:121–127
2. Fredriksson R, Lagerström MC, Lundin LG, Schiöth HB (2003) The G-protein-coupled receptors in the human genome form five main families. Phylogenetic analysis, paralogon groups, and fingerprints. Mol Pharmacol 63:1256–1272
3. Kawasawa Y, McKenzie LM, Hill DP, Bono H, Yanagisawa M et al (2003) G-protein-coupled receptor genes in the FANTOM2 database. Genome Res 13:1466–1477
4. DeVries ME, Kelvin AA, Xu L, Ran L, Robinson J, Kelvin DJ (2006) Defining the origins and evolution of the chemokine/chemokine receptor system. J Immunol 176:401–415
5. Rot A, von Andrian UH (2004) Chemokines in innate and adaptive host defense: basic chemokinese grammar for immune cells. Annu Rev Immunol 22:891–928
6. Li M, Ransohoff RM (2008) Multiple roles of chemokine CXCL12 in the central nervous system: a migration from immunology to neurobiology. Prog Neurobiol 84:116–131
7. Li M, Ransohoff RM (2009) The roles of chemokine CXCL12 in embryonic and brain tumor angiogenesis. Semin Cancer Biol 19: 111–115
8. Kokovay E, Goderie S, Wang Y, Lotz S, Lin G, Sun Y et al (2010) Adult SVZ lineage cells home to and leave the vascular niche via differential responses to SDF1/CXCR4 signaling. Cell Stem Cell 7:163–173
9. Corti S, Locatelli F, Papadimitriou D, Donadoni C, Del BR, Fortunato F et al (2005) Multipotentiality, homing properties, and pyramidal neurogenesis of CNS-derived LeX(ssea-1)+/CXCR4+ stem cells. FASEB J 19:1860–1872
10. Dziembowska M, Tham TN, Lau P, Vitry S, Lazarini F, Dubois-Dalcq M (2005) A role for CXCR4 signaling in survival and migration of neural and oligodendrocyte precursors. Glia 50:258–269
11. Gong X, He X, Qi L, Zuo H, Xie Z (2006) Stromal cell derived factor-1 acutely promotes neural progenitor cell proliferation in vitro by a mechanism involving the ERK1/2 and PI-3K signal pathways. Cell Biol Int 30:466–471
12. Peng H, Huang Y, Rose J, Erichsen D, Herek S, Fujii N et al (2004) Stromal cell-derived factor 1-mediated CXCR4 signaling in rat and human cortical neural progenitor cells. J Neurosci Res 76:35–50
13. Tran PB, Ren D, Veldhouse TJ, Miller RJ (2004) Chemokine receptors are expressed widely by embryonic and adult neural progenitor cells. J Neurosci Res 76:20–34
14. Wu Y, Peng H, Cui M, Whitney NP, Huang Y, Zheng JC (2009) CXCL12 increases human neural progenitor cell proliferation through Akt-1/FOXO3a signaling pathway. J Neurochem 109:1157–1167
15. Li M, Chang CJ, Lathia JD, Wang L, Pacenta HL, Cotleur A, Ransohoff RM (2011) Chemokine receptor CXCR4 signaling modulates the growth factor-induced cell cycle of self-renewing and multipotent neural progenitor cells. Glia 59:108–118

Chapter 5

Chemokine Receptor Interactions with Virus-Like Particles

Luigi Buonaguro, Maria Tagliamonte, and Maria Luisa Visciano

Abstract

Virus-like particles (VLPs) presenting conformational envelope proteins on their surface represent an invaluable tool to study molecular interactions between viruses and cellular receptors/co-receptors, eliminating biological risks associated with working with live native viruses. The availability of target cells expressing specific chemokine receptors facilitates the dissection of specific interactions between human immunodeficiency virus (HIV) viral envelope proteins and these receptors in the laboratory. Here, we describe a method to evaluate HIV-VLP binding to cellular chemokine co-receptors, by carboxyfluorescein succinimidyl ester labeling and cellular uptake.

Key words Cell binding, Chemokine receptors, CFSE labeling, Human immunodeficiency virus, Virus-like particles

1 Introduction

HIV-1 infection is initiated by the interaction of glycoprotein 120 (gp120), the external portion of the trimeric envelope (Env) complex, with the CD4 receptor expressed on the host cell surface. Following this, the co-receptor-binding site on gp120 is exposed, permitting surface-expressed chemokine co-receptor molecule (i.e., CCR5 or CXCR4) binding on CD4+ T lymphocytes. As a consequence of chemokine co-receptor binding to the gp120–CD4 complex, the transmembrane viral glycoprotein gp41 acquires the thermostable, six-helix bundle structure that apposes the viral and cellular membranes, resulting in the formation of fusion pores [1–8]. Characterizing the HIV-1 entry process and the structural changes of the human immunodeficiency virus (HIV) envelope glycoprotein gp120 after binding with its cognate cellular receptor and co-receptor may provide new opportunities to develop HIV-1-specific immunogens. This could provide further tools for vaccine development, as well as the design of targeted small molecular inhibitors of HIV-1 viral entry.

Astrid E. Cardona and Eroboghene E. Ubogu (eds.), *Chemokines: Methods and Protocols*, Methods in Molecular Biology, vol. 1013, DOI 10.1007/978-1-62703-426-5_5, © Springer Science+Business Media New York 2013

The use of specific binding assays is a necessary prerequisite to understand and investigate the determinants and kinetics of the viral binding process.

Recently, several groups have developed soluble trimers of the gp120–gp41 Env ectodomain (i.e., lacking the transmembrane, cytoplasmic domain, and gp140 component). These are considered to preserve or mimic the structure of functional Env trimeric complexes present on the native viral particles [9–12]. In addition to these soluble forms, HIV gp140 trimeric structures can be presented on nanoparticles including liposomes, inactivated viruses, and virus-like particles (VLPs) or pseudovirions, to closely mimic the native conformation in a membrane-bound state [13–16]. In particular, VLPs represent a complex structure based on viral capsid proteins which self-assemble into particles closely resembling immature native virus [17–20]. VLPs are replication as well as infection incompetent, lacking regulatory proteins as well as infectious genetic material [21–25]. Due to these characteristics, VLPs represent an ideal surrogate in lieu of infectious viruses for experimental manipulation. Moreover, owing to the biological and structural properties of HIV gp140 trimers expressed on the VLP surface, these can be utilized for detailed binding studies. This important tool facilitates the study of the native virus–host cell interaction that results in virus fusion to host cell membranes [26, 27].

HIV Pr55Gag-VLPs have been produced in our laboratory using a transient baculovirus expression system (BEVS) that expresses chimeric HIV gp140 envelope proteins (gp120 + gp41 ectodomain) [28]. In parallel to the BEVS strategy, we recently developed a stable double-transfected insect cell line for the constitutive production of HIV Pr55gag-VLPs that express trimeric forms of HIV gp140 protein on their surface [29]. This new approach improves the expression and extracellular secretion of VLPs circumventing many difficulties associated with the conventional baculovirus-infected cell approach [30–32]. Carboxyfluorescein succinimidyl ester (CFSE) is a fluorescent cell-staining dye, which can be retained within cells for long periods and is not transferred to adjacent cells. CFSE was originally developed as a fluorescent dye to label lymphocytes and track their migration within animals [33]. Subsequently, it has been used to monitor lymphocyte proliferation due to the progressive halving of CFSE fluorescence within daughter cells with each cell division [34]. Cells are preferentially labeled with the diacetate form of CFSE (CFDA-SE) which is activated into the fluorescent CFSE only after cellular internalization and cleavage by intracellular esterases. CFSE-labeled VLPs have been proven to be an efficient tool to study virus–cellular interactions and internalization mechanisms [27, 35, 36].

2 Materials

Prepare all solutions using analytical grade reagents and ultrapure water (prepared by purifying deionized water to reach an electrical resistivity of 18 MΩ cm at 25 °C) unless stated otherwise.

2.1 HIV-VLP Production and Labeling Assay

1. *Spodoptera frugiperda* (Sf9) insect cells.
2. TC100 medium supplemented with 10 % fetal calf serum (FCS) (e.g., from Gibco-BRL).
3. 6-well tissue culture plates.
4. Recombinant bacmid DNA.
5. Cellfectin® transfection reagents (recommended for Sf9 cells; from Gibco-BRL).
6. High Five™ cells, derived from *Trichoplusia ni* egg cell homogenates (e.g., Invitrogen).
7. Serum-free Sf-900 medium (e.g., Gibco-BRL).
8. 25 % sucrose solution in water.
9. Ultracentrifuge tubes (transparent, 30–50 mL capacity).
10. Tris–NaCl–EDTA (TNE) buffer: 10 mM Tris–HCl, pH 8.0 + 100 mM NaCl and 1 mM EDTA in water.
11. Carboxyfluorescein diacetate succinimidyl ester (CFDA-SE) (e.g., Vybrant® CFDA Cell Tracer Kit from Molecular Probes/ Invitrogen/Life Technologies: *see* **Note 1**). CFDA-SE is provided as lyophilized powder in individual vials, each containing 500 μg. These are stored at −20 °C for a maximum of 6 months. Immediately prior to use, CFDA-SE is dissolved in dimethylsulfoxide (DMSO) to a concentration of 10 mM and diluted in 1× phosphate-buffered saline (PBS) to the 10 μM working concentration. This solution should be used promptly and not stored for subsequent use.
12. Centrifugal filter device, e.g., Microcon® YM-30 (Millipore, Billerica, MA, USA) (*see* **Note 2**).

2.2 HIV-VLP Binding Assay

1. Human glioblastoma cell line, U87, has been generated to constitutively express the CD4 receptor along with CXCR4 or CCR5 co-receptors. This can be obtained from the AIDS Reagent Programs at National Institutes of Health (NIH) in the United States or National Institute for Biological Standards and Control (NIBSC) in the United Kingdom, generously donated by Dr. D. Littman. Culture U87 cells in Dulbecco's modified Eagle's medium (DMEM) supplemented with 10 % fetal bovine serum (FBS), 100 μg/mL penicillin/streptomycin, and 2 mM glutamine at 37 °C in an incubator with 95 % air, 5 % CO_2. Store aliquots in liquid nitrogen in a standard medium-based freezing buffer containing 10 % DMSO.

2. Cell culture medium: RPMI 1640 medium or DMEM, 2 mM l-glutamine (*see* **Note 3**), containing 1 % penicillin/streptomycin, 1 % nonessential amino acids, 1 % sodium pyruvate, 50 μM 2-mercaptoethanol, 200 μg/mL gentamicin with or without 1 μg/mL puromycin (*see* **Note 4**), and 10 % FCS (*see* **Note 5**). Each component is stored at −20 °C, except for 2-mercaptoethanol which is stored at 4 °C. The complete medium is sterilized by vacuum filtration through a 0.22 μm membrane and stored at 4 °C prior to use.
3. Wash buffer: 1× PBS.
4. 6-well tissue culture plates.
5. Centrifuge tubes (15 mL).

2.3 HIV-VLP Uptake Evaluation

1. Dissociation reagent: 2.5 % trypsin (10×) without phenol red.
2. Fixative: 2 % paraformaldehyde in 1× PBS.
3. Fluorescent microscope.
4. Flow cytometer with minimum requirement of 3-color detection, including fluorescein isothiocyanate (FITC) wavelength range.
5. Flow cytometry acquisition and analysis software, e.g., Becton Dickinson FACSCalibur™, CELLQuest™.

3 Methods

3.1 Insect Cell Cultures and HIV-VLP Production

1. Propagate Sf9 insect cells in TC100 medium supplemented with 10 % FCS in a tissue culture dish or flask. Expand in an incubator at 28 °C without CO_2. Harvest by standard methods as recommended by the vendor.
2. Seed 9×10^5 Sf9 cells in one well of a 6-well tissue culture plate. Transfect with 10–20 μg of recombinant bacmid DNA using the Cellfectin® method, according to the manufacturer's instructions.
3. For large-scale preparations of VLPs, High Five™ cells, derived from *Trichoplusia ni* egg cell homogenates, are recommended.
4. Propagate High Five™ cells in serum-free Sf-900 medium according to the vendor's instructions.
5. Infect with the recombinant baculovirus, released in the Sf9 cell culture supernatants, at a multiplicity of infection (m.o.i.) of 5. Propagate the insect cell line in an incubator at 28 °C without CO_2.
6. Collect supernatants from Sf9 or High Five™ cell cultures 4 days post transfection or infection.

7. Centrifuge at 2,000 × *g* for 15 min at 4 °C to pellet cells. Collect the supernatant to isolate VLPs.
8. Pellet VLPs by ultracentrifugation. Carefully layer the supernatant over a 25 % sucrose solution in a ratio of 10: 1 volume/volume, and spin at 100,000 × *g* for 75 min at 4 °C. Collect VLPs at the interface with a sterile transfer pipette.
9. Resuspend pellet in 40 mL of TNE buffer. Store VLPs at 4 °C for up to 2 weeks and at −20 °C for longer periods.

3.2 HIV-VLP Labeling Assay

1. Purify HIV-VLPs by ultracentrifugation over a 25 % sucrose cushion as described in Subheading 3.1, **step 8**, and quantify using the Bradford assay.
2. Reconstitute HIV-VLPs in 500 μL of buffer (e.g., 1× PBS), transfer to a 1.5 mL Eppendorf tube, and add 1 μL stock CFDA-SE (final concentration of 10 μM). Mix thoroughly by vortexing and incubate for 1 h at room temperature.
3. Load a maximum of 0.5 mL of the HIV-VLP/CFDA-SE mix into the centrifugal filtration device sample reservoir. Centrifuge at 14,000 × *g* for 10 min at room temperature using an appropriate benchtop centrifuge to remove the excess unbound CFDA-SE (*see* **Note 1**).
4. Place sample reservoir upside down in a new vial. Centrifuge at 1,000 × *g* for 3 min at room temperature to completely recover the labeled HIV-VLPs. The recovered labeled HIV-VLPs are further quantified by the Bradford assay.

3.3 HIV-VLP Binding Assay

1. Expand cryopreserved chemokine co-receptor-expressing cells (e.g., human glioblastoma U87 cells) in 3 mL of cell culture medium by plating 2×10^5 cells per mL in 6-well tissue culture-treated plates (total of 6×10^5 cells per well). Cells are cultured in a humidified incubator in an atmosphere of 95 % air and 5 % CO_2 until 90 % confluence.
2. Aspirate old medium and detach cells from the tissue culture plate using 3 mL of freshly made 1 % trypsin solution in 1× PBS. Once about 90 % of the cells are detached, harvest cells into 15 mL centrifuge tubes by repetitive tituration and neutralize trypsin with 7 mL regular growth medium. Pellet cells by centrifugation at 300 × *g* for 5 min at room temperature. Reconstitute cells at 1×10^6 cells per mL.
3. Determine cell viability using the trypan blue exclusion test (*see* **Note 6**).
4. Incubate 1×10^6 chemokine co-receptor-expressing cells with 10 μg/mL CFDA-SE-labeled HIV-VLPs in 1× PBS for 60 min at 37 °C in fresh 6-well tissue culture plates. Plates are kept in a humidified incubator in an atmosphere of 95 % air and 5 % CO_2.

5. Eliminate unbound VLPs by aspirating the 1× PBS containing labeled VLPs from the adherent cells and then wash the cells with 500 μL in 1× PBS.
6. Incubate cells bound with CFDA-SE-labeled HIV-VLPs with fresh RPMI 1640 medium at 37 °C for 2–24 h in a humidified incubator in an atmosphere of 95 % air and 5 % CO_2. At the end of incubation, cells are recovered from the 6-well plates by 1 % trypsin treatment, as described in item 2 above.

3.4 HIV-VLP Uptake Evaluation

At specific time points, chemokine co-receptor-expressing cells bound to CFDA-SE-labeled HIV-VLPs are observed under a fluorescence microscope in order to detect VLP uptake. These cells are then harvested, and washed with 1× PBS, and the amount of fluorescence quantified by flow cytometry.

1. Direct fluorescence microscopy: Cells treated with CFDA-SE-labeled HIV-VLPs are visualized directly in 6-well tissue culture plates using an inverted microscope with appropriate phase contrast and fluorescent filters at specific time points (Fig. 1).
2. Flow cytometry analysis: Aspirate culture medium from each well, and wash adherent cells twice with 500 μL 1× PBS/well.
3. Harvest cells with trypsin (2.5 % solution). Pre-warm trypsin to 37 °C in a water bath, and add 200 μL to each well and incubate at room temperature until all the cells are detached (as determined by direct visualization with an inverted phase microscope).

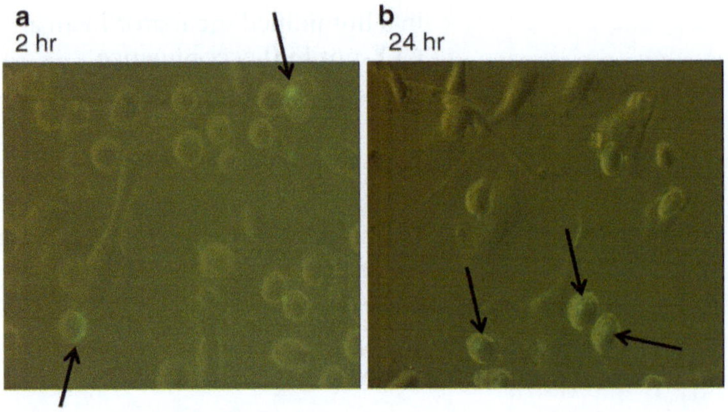

Fig. 1 Expression of CFDA-SE-labeled HIV-VLPs in chemokine co-receptor-expressing cells. Digital photomicrographs show CFDA-SE-labeled HIV-VLPs in harvested cells (*black arrows*) collected 2 and 24 h following incubation (**a** and **b**, respectively) as observed by direct fluorescence microscopy

4. Remove completely detached cells from the wells by gentle pipetting, and mix immediately with washing buffer (2 % FBS in 1× PBS) (*see* **Note 7**). Centrifuge at 300 × *g* for 5 min at 4 °C.
5. Resuspend pellet with 1 mL of 2 % paraformaldehyde in 1× PBS to fix the cells. Keep cells at 4 °C covered with aluminum foil prior to FACS analysis. Cells can be analyzed immediately or stored overnight in 1× PBS at 4 °C, following a brief wash by centrifugation.
6. Cells are gated according to forward- and side-scatter characteristics, with detection of HIV-VLP uptake based on gating on green fluorescence (FL1). Typically, 20,000 events are collected and data are analyzed and displayed as histograms using a flow cytometry analysis software program (Fig. 2, *see* **Note 8**).

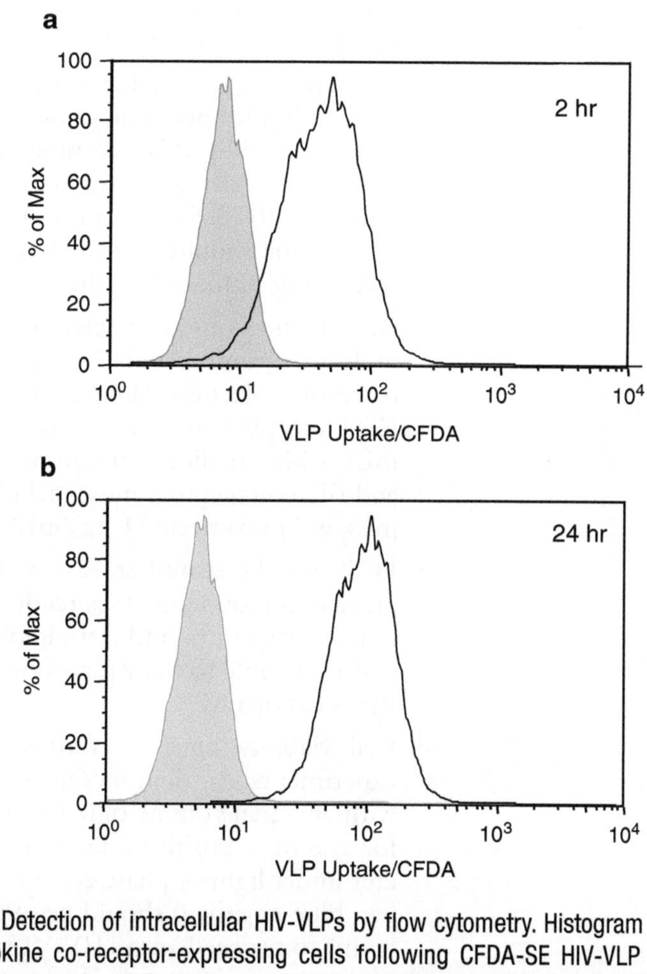

Fig. 2 Detection of intracellular HIV-VLPs by flow cytometry. Histogram plots of chemokine co-receptor-expressing cells following CFDA-SE HIV-VLP binding show progressive VLP uptake with time comparing 2 h (**a**) to 24 h (**b**). The *black curves* indicate intracellular HIV-VLP uptake, while the *shadowed curve* represents background fluorescence

4 Notes

1. The Vibrant CFDA-SE routinely used in our laboratory is obtained from Invitrogen/Life Technologies. However, the same product can be obtained from different vendors with similar results, as reported in literature.
2. Microcon YM-30 has a nominal molecular weight limit (NMWL) of 30,000 Da which ensures VLP retention on the filter membrane and the flow through of the unbound CFDA-SE which has a molecular weight of 557 Da. The optimal spin time using this device at 25 °C is approximately 10 min at 14,000 × *g*. Do not exceed this centrifugal force to avoid the outflow of labeled VLP through the filter and to prevent the damage to the device. Double centrifugation time if filtration is performed at 4 °C. If filter membrane becomes dry, add at least 10 μL of buffer (e.g., 1× PBS) to the sample reservoir to resuspend membrane-bound VLPs for recovery.
3. l-glutamine is very stable as a dry powder and as a frozen solution. In liquid media or stock solutions, l-glutamine degrades relatively rapidly. It is recommended to prepare the l-glutamine solution in cell culture-grade water and keep it in small aliquots at −20 °C for a maximum of a single thawing. Additional l-glutamine should be added every 1–2 weeks to the medium containing l-glutamine when stored at 4 °C.
4. Gentamicin is used to select cells that express CD4 receptor molecule; puromycin is used to select cells expressing the co-receptor molecules. Medium for culturing cells expressing only CD4 receptor molecule must include gentamicin (200 μg/mL), while medium for culture of cells expressing both CD4 and the co-receptors must include both gentamicin (200 μg/mL) and puromycin (1 μg/mL).
5. FCS must be stored at −20 °C to prevent degradation of its protein components. It is recommended to thaw it overnight in the refrigerator and not rapidly in a water bath at 37 °C. It is also possible to thaw it at room temperature, making sure to stir occasionally.
6. Cell viability must be verified before starting the binding experiments. An aliquot of detached cells (e.g., 20 μL) is mixed with an equal volume of 0.4 % trypan blue solution in 1× PBS for 2–3 min. Positive cells are enumerated with a hemacytometer under light or phase contrast microscopy. Viable cells (trypan blue negative) should be >90 % of the estimated total cell count to proceed with HIV-VLP binding.
7. FBS must be added to 1× PBS in order to neutralize trypsin.

8. Flow cytometer setup should be performed by experienced personnel to address issues that may arise during data acquisition that could significantly affect subsequent analysis of HIV-VLP uptake using this method.

References

1. Doms RW, Moore JP (2000) HIV-1 membrane fusion: targets of opportunity. J Cell Biol 151:F9–F14
2. Jones PL, Korte T, Blumenthal R (1998) Conformational changes in cell surface HIV-1 envelope glycoproteins are triggered by cooperation between cell surface CD4 and co-receptors. J Biol Chem 273:404–409
3. Sullivan N, Sun Y, Sattentau Q, Thali M, Wu D, Denisova G et al (1998) CD4-Induced conformational changes in the human immunodeficiency virus type 1 gp120 glycoprotein: consequences for virus entry and neutralization. J Virol 72:4694–4703
4. Sattentau QJ, Moore JP (1991) Conformational changes induced in the human immunodeficiency virus envelope glycoprotein by soluble CD4 binding. J Exp Med 174:407–415
5. Zhang W, Canziani G, Plugariu C, Wyatt R, Sodroski J, Sweet R et al (1999) Conformational changes of gp120 in epitopes near the CCR5 binding site are induced by CD4 and a CD4 miniprotein mimetic. Biochemistry 38: 9405–9416
6. Wyatt R, Sodroski J (1998) The HIV-1 envelope glycoproteins: fusogens, antigens, and immunogens. Science 280:1884–1888
7. Golding H, Zaitseva M, de Rosny E, King LR, Manischewitz J, Sidorov I et al (2002) Dissection of human immunodeficiency virus type 1 entry with neutralizing antibodies to gp41 fusion intermediates. J Virol 76:6780–6790
8. Eckert DM, Kim PS (2001) Mechanisms of viral membrane fusion and its inhibition. Annu Rev Biochem 70:777–810
9. Burton DR, Desrosiers RC, Doms RW, Koff WC, Kwong PD, Moore JP et al (2004) HIV vaccine design and the neutralizing antibody problem. Nat Immunol 5:233–236
10. Schulke N, Vesanen MS, Sanders RW, Zhu P, Lu M, Anselma DJ et al (2002) Oligomeric and conformational properties of a proteolytically mature, disulfide-stabilized human immunodeficiency virus type 1 gp140 envelope glycoprotein. J Virol 76:7760–7776
11. Srivastava IK, VanDorsten K, Vojtech L, Barnett SW, Stamatatos L (2003) Changes in the immunogenic properties of soluble gp140 human immunodeficiency virus envelope constructs upon partial deletion of the second hypervariable region. J Virol 77:2310–2320
12. Yang X, Lee J, Mahony EM, Kwong PD, Wyatt R, Sodroski J (2002) Highly stable trimers formed by human immunodeficiency virus type 1 envelope glycoproteins fused with the trimeric motif of T4 bacteriophage fibritin. J Virol 76:4634–4642
13. Grundner C, Mirzabekov T, Sodroski J, Wyatt R (2002) Solid-phase proteoliposomes containing human immunodeficiency virus envelope glycoproteins. J Virol 76:3511–3521
14. Race E, Frezza P, Stephens DM, Davis D, Polyanskaya N, Cranage M et al (1995) An experimental chemically inactivated HIV-1 vaccine induces antibodies that neutralize homologous and heterologous viruses. Vaccine 13:54–60
15. Buonaguro L, Buonaguro FM, Tornesello ML, Mantas D, Beth-Giraldo E, Wagner R et al (2001) High efficient production of Pr55gag Virus-like Particles expressing multiple HIV-1 epitopes, including a gp120 protein derived from an Ugandan HIV-1 isolate of subtype A. Antivir Res 49:35–47
16. Crooks ET, Moore PL, Franti M, Cayanan CS, Zhu P, Jiang P et al (2007) A comparative immunogenicity study of HIV-1 virus-like particles bearing various forms of envelope proteins, particles bearing no envelope and soluble monomeric gp120. Virology 366: 245–262
17. Gheysen D, Jacobs E, de Foresta F, Thiriart C, Francotte M, Thines D et al (1989) Assembly and release of HIV-1 precursor $Pr55^{gag}$ virus-like particles from recombinant baculovirus-infected insect cells. Cell 59: 103–112
18. Delchambre M, Gheysen D, Thines D, Thiriart C, Jacobs E, Verdin E et al (1989) The GAG precursor of simian immunodeficiency virus assembles into virus-like particles. EMBO J 8:2653–2660
19. Miyanohara A, Imamura T, Araki M, Sugawara K, Ohtomo N, Matsubara K (1986) Expression of hepatitis B virus core antigen gene in Saccharomyces cerevisiae: synthesis of two polypeptides translated from different initiation codons. J Virol 59:176–180

20. Kirnbauer R, Booy F, Cheng N, Lowy DR, Schiller JT (1992) Papillomavirus L1 major capsid protein self-assembles into virus-like particles that are highly immunogenic. Proc Natl Acad Sci USA 89:12180–12184
21. Takamura S, Niikura M, Li TC, Takeda N, Kusagawa S, Takebe Y et al (2004) DNA vaccine-encapsulated virus-like particles derived from an orally transmissible virus stimulate mucosal and systemic immune responses by oral administration. Gene Ther 11:628–635
22. Malboeuf CM, Simon DA, Lee YE, Lankes HA, Dewhurst S, Frelinger JG et al (2007) Human papillomavirus-like particles mediate functional delivery of plasmid DNA to antigen presenting cells in vivo. Vaccine 25: 3270–3276
23. Touze A, Coursaget P (1998) In vitro gene transfer using human papillomavirus-like particles. Nucleic Acids Res 26:1317–1323
24. Xu YF, Zhang YQ, Xu XM, Song GX (2006) Papillomavirus virus-like particles as vehicles for the delivery of epitopes or genes. Arch Virol 151:2133–2148
25. Buonaguro L, Buonaguro FM, Tornesello ML (2010) Virus-like particles as particulate vaccines. Curr HIV Res 8:299–309
26. Triyatni M, Saunier B, Maruvada P, Davis AR, Ulianich L, Heller T et al (2002) Interaction of hepatitis C virus-like particles and cells: a model system for studying viral binding and entry. J Virol 76:9335–9344
27. Buonaguro L, Tornesello ML, Tagliamonte M, Gallo RC, Wang LX, Kamin-Lewis R et al (2006) Baculovirus-derived human immunodeficiency virus type 1 virus-like particles activate dendritic cells and induce ex vivo T-cell responses. J Virol 80:9134–9143
28. Visciano ML, Diomede L, Tagliamonte M, Tornesello ML, Asti V, Bomsel M et al (2011) Generation of HIV-1 virus-like particles expressing different HIV-1 glycoproteins. Vaccine 29:4903–4912
29. Tagliamonte M, Tornesello ML, Buonaguro FM, Buonaguro L (2011) Conformational HIV-1 envelope on particulate structures: a tool for chemokine coreceptor binding studies. J Transl Med 9(Suppl 1):S1
30. Harrison RL, Jarvis DL (2007) Transforming lepidopteran insect cells for continuous recombinant protein expression. Methods Mol Biol 388:299–316
31. Jarvis DL, Fleming JA, Kovacs GR, Summers MD, Guarino LA (1990) Use of early baculovirus promoters for continuous expression and efficient processing of foreign gene products in stably transformed lepidopteran cells. Biotechnology (NY) 8:950–955
32. Kato T, Murata T, Usui T, Park EY (2004) Comparative analysis of GFP(UV)-beta1,3-N-acetylglucosaminyltransferase 2 production in two insect-cell-based expression systems. Protein Expr Purif 35:54–61
33. Weston SA, Parish CR (1990) New fluorescent dyes for lymphocyte migration studies. Analysis by flow cytometry and fluorescence microscopy. J Immunol Methods 133:87–97
34. Lyons AB, Parish CR (1994) Determination of lymphocyte division by flow cytometry. J Immunol Methods 171:131–137
35. Drobni P, Mistry N, McMillan N, Evander M (2003) Carboxy-fluorescein diacetate, succinimidyl ester labeled papillomavirus virus-like particles fluoresce after internalization and interact with heparan sulfate for binding and entry. Virology 310:163–172
36. Bergsdorf C, Beyer C, Umansky V, Werr M, Sapp M (2003) Highly efficient transport of carboxyfluorescein diacetate succinimidyl ester into COS7 cells using human papillomavirus-like particles. FEBS Lett 536: 120–124

Chapter 6

Chemokine Receptor Antagonist Development

Alexandre Garin, Zoë Johnson, Aurelie Hermant, Fanny Beltran, Yann Ratinaud, Alexandra Michel, Sonja Krohn, Marilène Gaudet, Susanna Carboni, Hong Ji, Marc Missotten, Olivier Leger, Christine Power, and Amanda Proudfoot

Abstract

This chapter describes assays that focus on the characterization of compounds identified in high-throughput screening campaigns, and the subsequent medicinal chemistry programs. They cover methods to determine potency in buffer, the effect of whole blood on the compounds' activity and finally the pharmacokinetic (PK)/pharmacodynamic (PD) relationship of the compounds in a rodent species.

Key words Chemokine, Chemokine receptor antagonists, Glycosaminoglycans, Receptor coverage ex vivo

1 Introduction

Over the past 25 years, research studies have established the biological importance of the cytokine subfamily, *chemo*attractant cyto*kines* or chemokines: molecules with the principal role to direct cellular migration. Thus, chemokines and their receptors play a major role in leukocyte homing and in the recruitment of leukocytes to inflammatory sites. However, their function is not limited to their role as chemoattractants, as they are involved in numerous biological processes ranging from development, hematopoiesis, angiogenesis, cellular activation and proliferation, neuromodulation, tumor growth, and metastasis. Worldwide interest in the chemokine field surged dramatically early in 1996, with the finding that certain chemokine receptors were the elusive co-receptors, required along with CD4, for human immunodeficiency virus (HIV) infection. Since then, these molecules have been described as broader mediators in infection.

Astrid E. Cardona and Eroboghene E. Ubogu (eds.), *Chemokines: Methods and Protocols*, Methods in Molecular Biology, vol. 1013, DOI 10.1007/978-1-62703-426-5_6, © Springer Science+Business Media New York 2013

Chemokines are a family of small proteins, of which approximately 50 have been identified in humans. They are unique in the cytokine family as they exert their effects through a family of G-protein-coupled receptors (GPCRs) that are seven-transmembrane spanners (7TM). To date, 19 receptors have been identified, and one can immediately see that the system is promiscuous in that there are few receptors that only bind one ligand, and furthermore, certain ligands bind to more than one receptor. The discovery of this class of receptors caused great excitement in the pharmaceutical industry, since 7TM receptors are one of the cornerstones of drug discovery. There have been intense efforts to develop inhibitors of the chemokine system, which is reflected by the large number of small molecule inhibitor patents [1]. However, although there have been several clinical trials in several different diseases, the results have been disappointing [2]. The promiscuity in ligand–receptor interactions, in addition to the large size of the ligands, around 8,000 Da compared to <200 Da for ligands such as serotonin and histamine represented a great challenge for the medicinal chemists. Moreover, certain receptors have proved to be more difficult to find inhibitors for than others, for example the number of patents for compounds binding to the interleukin (IL)-8 receptors CXCR1 and CXCR2 differs considerably. Understanding chemokine ligand and receptor function depends on several assays, many of which are also used for drug discovery. This chapter will describe assays that have been recently developed in our laboratory, complementing those previously described [3].

1.1 Chemotaxis for Drug Development

Chemotaxis is the principal activity mediated by chemokines. Upon stimulation with a chemokine, cells that express the cognate receptor on their plasma membranes migrate towards the directional signal provided by the chemokine. A bell-shaped concentration-effect or dose-response curve is a common feature of chemotaxis assays in vitro. Chemotaxis assays can be performed in many ways (e.g., Boyden Chamber, Zigmond chamber, Dunn chamber, Agar-plate techniques), but all are of low throughput. On the contrary, multi-well chambers (usually containing 96 wells) allow the testing of two to three compounds per plate and are amenable for high-throughput screening. We routinely perform measurements in triplicate using threefold dilutions covering the dose range of 10^{-6}–10^{-12} M in 96-well plates, allowing only two assays per plate. Obviously the number of dilutions can be reduced, but the number of assays that can be performed is limited to approximately 10 per day by an experienced worker. We have introduced automation of certain steps which results in an increased number of compounds tested per day to 80 compounds with ten dose–response points in triplicate or 720 compounds at a single concentration in duplicate. This facilitates improved reproducibility and robustness of the assay.

The basic approach of this assay is to place the chemokine in the lower chamber of the chemotaxis system, allowing diffusion through the transwell membrane pores to activate the cells in the upper chamber. Quantification of the migrated cells can be performed by several methods, and here we describe an indirect quantification using a fluorescent DNA staining dye. We consider chemotaxis assays as the method of choice to determine potency and activity of antagonists or neutralizers of chemokines or their receptors. We recommend that compounds which are advanced in the optimization process are then tested for potency in serum. The activity of a given chemokine in chemotaxis buffer might not reflect its real activity in vivo, since several chemokines are known to be processed by proteases in the blood circulation [4]. Secondly, chemokine receptor inhibitors often interact with plasma proteins and thus may lose potency in comparison to their activity in buffer.

1.2 Cellular Dielectric Spectroscopy

The CellKey™ system is a label-free technology that uses cellular dielectric spectroscopy (CDS) to measure changes in cell-substrate adherence, changes in cell shape and volume, and changes in cell–cell interactions in live cells. Unlike traditional technologies that measure only discrete intracellular events, such as Ca^{2+} flux or cyclic adenosine monophosphate (cAMP) fluctuation, the CellKey™ system measures the integrated response of the whole cell to receptor activation. The CellKey™ system measures the changes in cell adherence, cell shape and volume, and cell–cell interactions in the complex impedance (delta Z or dZ) of a cell monolayer that occurs in response to receptor stimulation. Impedance (Z) is defined by the ratio of voltage to current as described by Ohm's law ($Z = V/I$). Cells are seeded onto a custom CellKey™ microplate that contains electrodes at the bottom of each well. The voltage is applied through different frequencies to record both extracellular and intracellular currents leading to a unique ligand-induced response depending on the receptor coupling: Gi, Gs, or Gq for GPCRs. This technology enables comprehensive pharmacological evaluation of both endogenous and recombinant transfected cells, without the need for any cell modification, tagged receptor, labeled ligands, or specialized reagents. The response of several types of receptors can be monitored including GPCRs, tyrosine kinase receptors (TKRs), and other cellular targets. Chemokine receptor antagonists can be characterized using this system; as an example in this section, the protocol will be based on CCR2 and its endogenous ligand CCL2.

1.3 β-Arrestin

In response to a stimulus, GPCRs activate heterotrimeric G proteins. The translocation and binding of β-arrestin to a phosphorylated GPCR is a necessary event leading to receptor desensitization and internalization after receptor activation by a chemokine. The binding of β-arrestin to the GPCR is proximal to receptor activation and occurs with both G protein-dependent and -independent signaling. Detection of the interaction between activated receptor and β-arrestin therefore offers the opportunity for a universal GPCR signaling assay format. Quantitatively examining the complex relationship that exists between G-protein and β-arrestin signaling determines both efficacy and potential side effects of GPCR-targeted drugs. In this section, we describe the use of the PathHunter eXpress β-arrestin assay developed and commercialized by DiscoveRx Corporation. The principle of this technology is enzyme fragment complementation. In this system, the β-galactosidase (β-gal) enzyme is split into two inactive fragments. The larger portion of β-gal is fused to the C-terminus of β-arrestin and the smaller fragment is expressed as a fusion to the C-terminal of the GPCR of interest. Upon activation, the interaction of the GPCR and β-arrestin brings the two β-gal fragments together resulting in enzyme fragment complementation and the formation of a functional enzyme capable of hydrolyzing substrate and generating a chemiluminescent signal.

1.4 ELISA-Based Heparin Binding Assay

Chemokines are known to interact with glycosaminoglycans (GAGs). This chemokine–GAG interaction has been shown to be an absolute requirement for cellular recruitment in vivo [5]. The ELISA-based heparin binding assay is a useful tool to investigate chemokine–GAG interactions (Fig. 1). Using heparin binding plates (initially supplied as Plasso EpranEx™ plates), GAGs can be adsorbed onto a specially prepared plate surface in a functionally active form by retaining their protein-binding characteristics [6]. However it should be noted that certain chemokines bind equally well to the bare plate as to the heparin-coated plate, so this must be taken into account when using this assay. A wide range of GAG

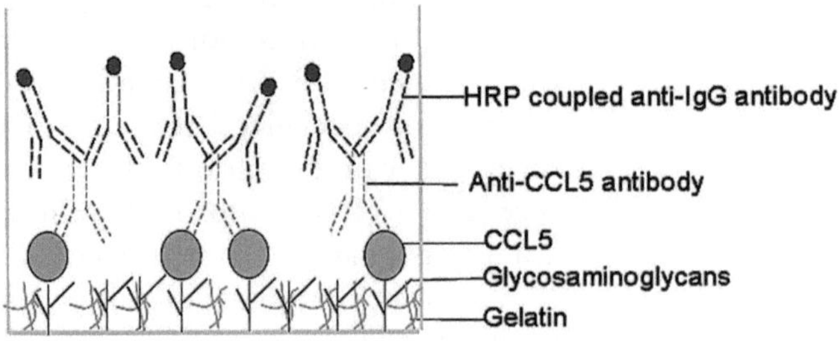

Fig. 1 ELISA-based heparin binding assay. The heparin/GAGs are bound onto the heparin binding plate. The surface of the wells is blocked with gelatin. The plate is incubated with the chemokine of interest, in this example CCL5. Bound CCL5 is detected using anti-CCL5 antibodies and visualized using HRP-coupled secondary antibodies

preparations (e.g., heparin, heparan sulfate, dermatan sulfate, or chondroitin sulfate) can be immobilized onto the heparin binding plate, which enables the determination of the chemokine affinity to the GAG of interest. In this section we describe the use of heparin, and the chemokine CCL5 as an example.

1.5 Chemokine-Induced Erk Phosphorylation in Mouse, Human, and Cynomolgus Whole Blood

The binding of chemokines to chemokine receptors induces activation of multiple intracellular signal transduction pathways, including mitogen-associated protein kinase (MAPK)/Erk1/2, phosphatidylinositol-3-kinase (PI3K)/Akt, and the transcription factor nuclear factor (NF)κB, eventually leading to transcription of specific target genes whose protein products are involved in regulating cell survival, proliferation, migration, and cell cycle progression. These pathways are complex, with extensive interconnections by synergistic and antagonistic cross talks; therefore a signaling event as well as the overall outcomes are often cell, time, and environmental milieu dependent. In this chapter, we describe a whole blood method using flow cytometry, combining phospho-specific antibodies to cell markers, to assess Erk phosphorylation at single cell level in the relevant physiological environment, blood. We have used B cell stimulation by CXCL13 as the example. This protocol, with minor adaptations, is applicable to mouse, human, and cynomolgus blood samples, which potentially can be used to monitor PD effects of receptor antagonists. It can be used to study any required receptor on other leukocyte subpopulations.

1.6 Alexa-Chemokine Binding in Human, Mouse, Dog, and Cynomolgus Blood

Alexa-labeled chemokines are chemokines chemically engineered for measurement using wavelengths detectable by flow cytometry instruments whilst still retaining their full receptor-binding capacities. Such reagents can be used to stain the population that binds a given chemokine similarity to labeled antibodies. As this staining is quantitative, Alexa-labeled chemokines can be used in competition assays to monitor the efficacy of chemokine receptor antagonist or antibody binding inhibition. Given that many Alexa-labeled chemokines also bind to orthologous receptors (i.e., from different species), this technique can also be used to address drug cross-reactivity with other species. It is important to perform experiments in relevant biological fluids, and the described assay was developed in whole blood, to test the potency and cross-reactivity of CCR2 antagonists.

1.7 Thioglycollate Peritoneal Recruitment with Receptor Coverage

Correlating the PD effects of a drug with drug concentration and receptor coverage is important in preclinical drug discovery and clinical trial design. Here we describe a simple peritoneal recruitment method that can be used to correlate biological activity of a chemokine antagonist in vivo with measured chemokine receptor occupancy ex vivo using flow cytometry. Intraperitoneal injection of thioglycollate broth has been widely used to induce acute inflammation in the peritoneal cavity of rodents, with a predominance of monocyte infiltration occurring at 24 h. Flow cytometry is a powerful tool to characterize the cellular composition of inflammatory infiltrates, enabling quantification of the effect of a receptor antagonist on specific cellular subsets. Combining these methods allows the quantification of CCR2 and CCR5 occupancy by a known, dual-CCR2/CCR5 antagonist in order to correlate compound effect at blocking cell recruitment with receptor coverage. The amount of receptors occupied by the antagonist will be reflected by a proportional loss of Alexa-647-conjugated chemokine binding, and measured by a decrease in signal intensity using flow cytometry.

2 Materials

2.1 Chemotaxis for Drug Development

1. 96-well plastic V-bottom plate.
2. 96-well chemotaxis plates with 5 μM pore membranes (e.g., Neuro Probe Inc.®).
3. 96-well flat-bottom black plates (e.g., Corning).
4. Funnel adaptors (e.g., Neuro Probe Inc.®).
5. Filtered human or mouse serum.
6. SYBR green (e.g., Invitrogen).
7. Recombinant chemokine of interest (produced in house or commercially available from several sources such as Peprotech, R&D systems, Almac, etc.) (*see* **Note 1**).
8. Primary cells or cell lines expressing the relevant chemokine receptor.
9. Culture medium: RPMI 1640 medium, containing 10 % heat-inactivated fetal calf serum (FCS), 1 % l-glutamine, and 1 % penicillin/streptomycin.
10. Chemotaxis medium: RPMI 1640 medium without phenol red, containing 5 % heat-inactivated FCS.
11. Dimethyl sulfoxide (DMSO).
12. Butyrate.

2.2 Cellular Dielectric Spectroscopy

The following description is based on a chemokine receptor coupled to Gi subunit development transfected in Chinese hamster ovary (CHO) cells, using CCR2 as the example.

1. CHO cells transfected with the chemokine receptor of interest, for example as described below for CCR2.
2. CellKey™ assay buffer: Hank's balanced salt solution (HBSS), containing 20 mM HEPES, 0.1 % bovine serum albumin (BSA), with or without 1 % DMSO adjusted to pH 7.4.

The CellKey™ buffer containing 1 % DMSO is pre-warmed to 37 °C to wash cells.

3. Culture medium: Ham's F12 medium containing 10 % FCS, 1 % penicillin/streptomycin, sterile filtered using a 0.2 μM pore vacuum filtration device.
4. T225 Flask for cell culture.
5. Phosphate-buffered saline (PBS) buffer.
6. Geneticin®.
7. Buffer A for cell counting: Proprietary lysis/disaggregation buffer (Chemometec).
8. Buffer B for cell counting: Proprietary stabilizing buffer (Chemometec).
9. Inhibitor compounds reconstituted at 10 mM in 100 % DMSO.
10. Recombinant CCL2 (either produced in house or commercially available for several sources for example PeproTech, R&D systems, etc.) diluted to 100 μM in distilled H_2O.
11. CellKey™ Standard 96 W Microplates.
12. Polypropylene plates 96-well format.
13. CellKey™ instrument (MDS Sciex).
14. Class II Safety hood.
15. Incubator for cell culture: Maintained at 30 and 37 °C with 5 % CO_2, 95 % air.
16. Pipettes and appropriate tips.
17. CiBi TipTrays 96 (e.g., from Cybio).
18. Cell counting instrument such as Nucleocounter (Chemometec).

2.3 β-Arrestin

1. More than 150 human and mouse Gi-, Gq-, and Gs-coupled GPCRs are commercially available. We recommend the DiscoveRx Off-the-shelf complete kit (DiscoveRx Corporation) containing:
   (a) PathHunter™ express receptor-specific β-arrestin cells.
   (b) PathHunter detection reagents.
   (c) Optimized cell culture medium.
   (d) 96-well tissue culture-treated plates.
2. Optimized step-by-step protocol for 96- or 384-well format.

2.4 ELISA-Based Heparin Binding Assay

1. Standard assay buffer (SAB): 100 mM NaCl, 50 mM NaOAc, 0.2 % (v/v) Tween-20 in distilled H_2O, adjusted to pH 7.2 by adding NaOH.
2. PBS buffer.

3. Heparin solution: 50 μg/ml heparin sodium salt (e.g., Sigma-Aldrich) in PBS buffer (*see* **Note 2**). Prepare the solution on the day of the experiment.
4. Blocking solution: 0.2 % gelatin from bovine skin in PBS buffer (*see* **Note 3**). Prepare the solution on the day of the experiment.
5. Heparin binding plate (e.g., BD Biosciences).
6. Sealers (e.g., Perkin Elmer).
7. 96-well round-bottom cell culture cluster plates, with lid (e.g., Costar).
8. Human CCL5/RANTES produced in-house [7], or obtained from commercial sources.
9. Polyclonal anti-human CCL5/RANTES antibody.
10. Horseradish peroxidase (HRP)-conjugated donkey anti-goat IgG antibodies (e.g., Santa Cruz Biotechnology, Inc.).
11. HRP substrate tetramethylbenzidine (TMB) solution for ELISA (e.g., Uptima).
12. Sulfuric acid (H_2SO_4) (1 mol/l or 2 N). Store safely in a safety storage cabinet at room temperature with air exchange (*see* **Note 4**).

2.5 Chemokine-Induced Erk Phosphorylation in Mouse, Human, and Cynomolgus Whole Blood

Female C57BL/6 mice weighing 20–22 g are used for this representative experiment.

1. 4 % FCS in 1× PBS.
2. Sterile distilled H_2O.
3. Murine CXCL13 (e.g., from PeproTech).
4. Rabbit anti-phospho-MAPK (Erk) (e.g., from Cell Signaling).
5. Rat anti-mouse CD45R/B220-PerCP-Cy5.5 monoclonal antibody [mAb] (e.g., BD Pharmingen).
6. Alexa Fluor® 647 donkey anti-rabbit IgG antibody (e.g., from Invitrogen).
7. Disposable capillary pipettes, minicaps, end-to-end (e.g., Hirschmann Laborgeräte).
8. Lithium-heparin microtubes for blood samples.
9. Falcon 5 and 15 ml polypropylene tubes.
10. Centrifuge.
11. Water bath.
12. 96-well plate, V-shaped bottom.
13. Flow cytometry instrument equipped with a 633 helium-neon laser (red laser, e.g., FACSCalibur from BD biosciences).

14. Flow cytometry analysis software, e.g., FlowJo from Treestar, version 7.2.5.
15. CellQuest Pro software (Becton Dickinson).

2.6 Alexa-Chemokine Binding in Human, Mouse, Dog, and Cynomolgus Monkey Blood

1. Fluorescence-activated cell sorting (FACS) buffer: 1× PBS containing 0.1 % BSA and 0.01 % sodium azide.
2. Ice-cold fixative: 0.25 % formaldehyde in 1× PBS.
3. Lysis solution: 0.15 M NH_4Cl, 10 mM $NaHCO_2$, and 1 mM EDTA in distilled H_2O.
4. Alexa Fluor 647-CCL2 (e.g., from Almac Sciences).
5. 96-well assay black plate, 2 ml (e.g., Costar).

2.7 Thioglycollate Peritoneal Recruitment with Receptor Coverage

Experiments are performed using female C57BL/6 mice weighing 20–22 g.

1. Fluid thioglycollate medium—ready to use (e.g., Becton Dickinson). The ready-to-use thioglycollate is delivered as small vial with 20 ml of medium. It must be kept in the fridge at 4 °C and brought to room temperature before use. Medium should be checked for any impurities before use—it should be translucent with no cloudy deposits. The medium must be kept sterile and discard each opened aliquot after use.
2. Sterile 0.9 % NaCl in distilled H_2O.
3. PBS buffer.
4. PBS/EDTA: PBS buffer containing 1 mM EDTA.
5. Carboxymethylcellulose (CMC) sodium salt.
6. Tween-20.
7. Hydroxypropylmethylcellulose(HPMC)(e.g.,Methocel™K4M CR from Colorcon).
8. FACS buffer: PBS containing 0.1 % BSA and 0.01 % sodium azide.
9. Ice-cold fixative: 0.25 % formaldehyde in 1× PBS, kept on ice.
10. Lysis solution: 0.15 M NH_4Cl, 10 mM sodium bicarbonate, and 1 mM EDTA in distilled H_2O.
11. Alexa Fluor 647-CCL2 (e.g., Almac Sciences).
12. Alexa Fluor 647-CCL4 (e.g., Almac Sciences).
13. Mettler Toledo scale, e.g., type PG6002-5, for weighing rodents.
14. 1 ml Sterile syringes, e.g., Becton Dickinson Plastipak™.
15. Sterile needles, e.g., Becton Dickinson Microlance™ 25 G (0.5 × 16 mm) for intraperitoneal (i.p.) injections.
16. Sterile needles, e.g., BD Microlance™ 20 G (0.9 × 40 mm) for peritoneal lavage.

17. Falcon 14 and 15 ml polypropylene tubes.
18. 1 ml pipette with plastic tips.
19. Dissection material (forceps, scissors, dissection plate).
20. Centrifuge, e.g., Megafuge 1.0R, Heraeus Kendro Laboratories.
21. 2 ml Eppendorf tubes.
22. 6-in. plastic transfer pipette, non-sterile.
23. Feeding oral needles (e.g., from Fine Science Tools).
24. Veterinary equipment Minerve (51310 Esternay France) or Barré-hypnos anesthesia table for anesthesia.
25. Hematology analyzer, e.g., Beckman coulter AcT 5 Diff model.
26. 96-well assay block plate, 2 ml.

3 Methods

3.1 Chemotaxis for Drug Development

3.1.1 Preparation of the Chemotaxis Plates

1. Prepare cells, count them (*see* **Note 5**), and resuspend them at the required concentration (*see* **Note 6**) in chemotaxis medium or 100 % serum.
2. For a dose–response experiment in chemotaxis medium, prepare serial dilutions of the chemokine of interest in a 96-well round-bottom plate (*see* **Note 1**). Dilute the chemokine to a concentration of 1 μM in chemotaxis medium and prepare threefold dilutions covering a dose range of 10^{-6} to 10^{-10} (*see* **Note 7**).
3. For dose–response experiments in serum, prepare a dilution series of the chemokine of interest in a 96-well round-bottom plate. Dilute the chemokine to a concentration of 5 μM in chemotaxis medium and prepare threefold dilutions covering a dose range of 10^{-6} to 10^{-10} (*see* **Note 8**).
4. Transfer 31 μl of each chemokine dilution in triplicate into the lower wells (rows A, B, and C for instance).
5. Cover the plate with the porous membrane.
6. Add 20 μl of cells to each of the 96 spots on the membrane without touching the membrane.
7. Put the plastic cover on the plate.

3.1.2 Cell Chemotaxis and Migration Measurement

1. Incubate the plate at 37 °C in 5 % CO_2 and 95 % air for the required time (*see* **Note 8**).
2. Wash the outer surface of the upper membrane with 20 ml PBS to remove the cells that have not migrated.
3. Carefully remove the membrane.

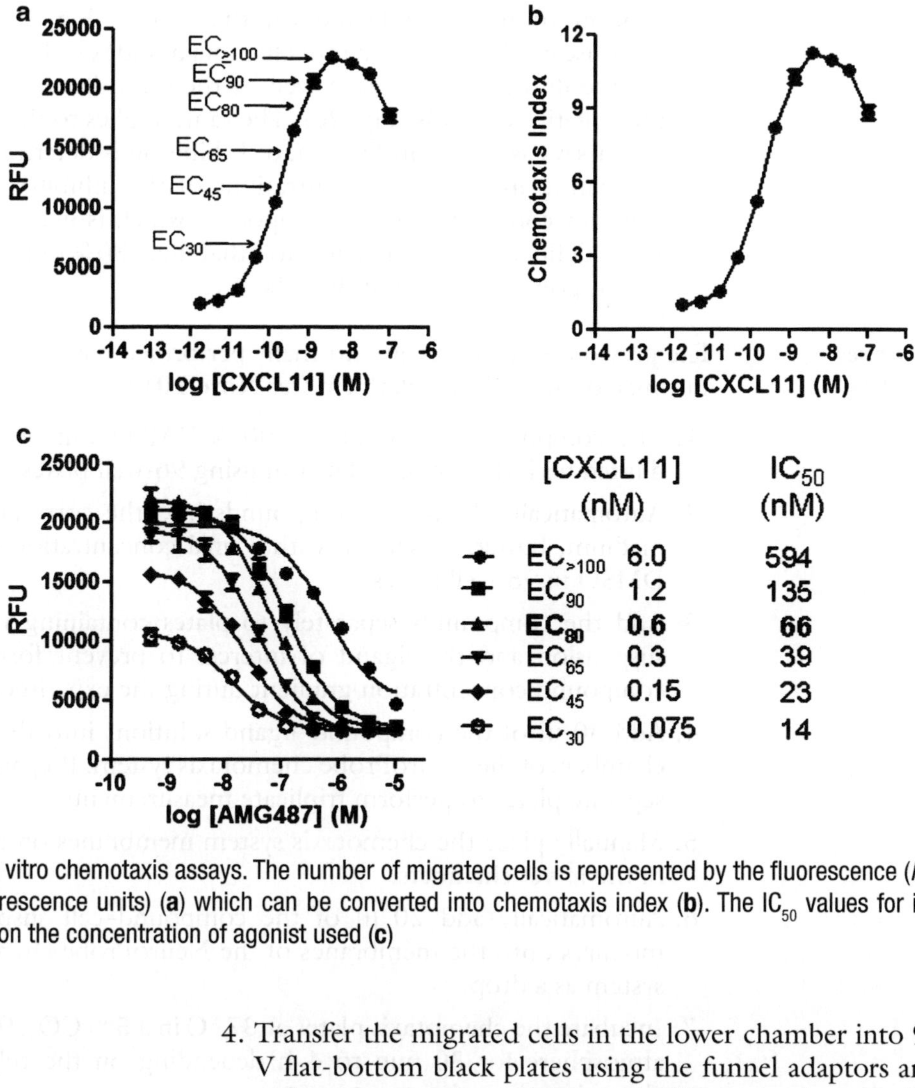

Fig. 2 In vitro chemotaxis assays. The number of migrated cells is represented by the fluorescence (*RFU* relative fluorescence units) (**a**) which can be converted into chemotaxis index (**b**). The IC_{50} values for inhibition depend on the concentration of agonist used (**c**)

4. Transfer the migrated cells in the lower chamber into 96-well V flat-bottom black plates using the funnel adaptors and centrifuge for 5 min at 570×*g* (2,000 rpm using the Heraeus Megafuge 16-R centrifuge) at room temperature.
5. Thaw the SYBR green solution (*see* **Note 9**).
6. Prepare the fluorescent solution. For 1 plate add 2 ml 1 % Triton-X-100 and 2 μl SYBR green to 18 ml H_2O.
7. Add 200 μl fluorescent solution per well and mix well to ensure total cell lysis accomplished by the detergent Triton-X-100 that facilitates the intracellular penetration of the SYBR green DNA dye.
8. Read the plate in a fluorescent plate reader.
9. Analysis can be carried out with the GraphPad Prism software. Chemotaxis responses in vitro generally generate a bell-shaped curve (Fig. 2). The data are plotted as fluorescence corresponding to the number of migrated cells on the *Y*-axis and the

concentration of the ligand on the *X*-axis (Fig. 2a). The response can be converted into chemotaxis index (CI) by dividing the fluorescence of each well by the mean fluorescence of the negative controls (Fig. 2c). The same applies to the inhibition assays as shown in Fig. 2b, d. Inhibition is determined by incubating increasing concentrations of the inhibitor with a constant concentration of the agonist, which is generally the EC_{80}, defined as the concentration that gives 80 % of the maximal response, as shown in Fig. 2a, c.

3.1.3 Automated Chemotaxis Assay

The previous steps which are normally performed manually can be automated on a robotic platform (*see* **Note 10**).

1. The compounds dissolved in 100 % DMSO can be serially diluted with the robotic platform using 96-well plates.
2. Automatically dilute the compounds into the corresponding medium, buffer, or serum, with a final concentration of 2 % DMSO in 96-well plates.
3. Add the compounds separately to plates containing the cell suspension and the ligand of interest to prevent forming a compound concentration gradient during the experiment.
4. Add 30 μl of the compound–ligand solutions into the lower chambers of the NeuroProbe chemotaxis system. Prepare three separate plates to perform triplicate measurements.
5. Manually place the chemotaxis system membranes on the top of the lower chambers.
6. Automatically add 20 μl of the compound–cell suspension mixtures onto the membranes of the NeuroProbe chemotaxis system as a drop.
7. Incubate the chemotaxis plates at 37 °C in a 5 % CO_2, 95 % air atmosphere for 30 min to 4 h depending on the cell type, chemokine ligand, or receptor.
8. Manually remove the drop of cells and the membranes from the chemotaxis system and transfer the cells with a funnel plate in 96-well flat-bottom black plates by centrifugation at 450 × *g* for 5 min.
9. Add 200 μl of the SYBR green solution (see Subheading 3.1.2, **step 6**) into each well of each plate using the robot and thoroughly mix.
10. Measure fluorescence by a fluorometer. Analyze data as described in Subheading 3.1.2, **step 9**.
11. The data normally produce a bell-shaped curve (Fig. 2a). The maximum effector concentration can be deduced from the peak of the curve.
12. To analyze chemotaxis inhibition, process data as described in Subheading 3.1.2, **step 9**.

3.2 Cellular Dielectric Spectroscopy

All the following steps are performed under sterile conditions. Medium is filtered and kept at 4 °C. CHO cells are harvested when they are ~100 % confluent and new flasks should not be seeded below 5 % confluence. Cells should be used prior to passage 25.

3.2.1 Day 1: Cell Culture and Assay Plate Preparation for Adherent Cells

1. Wash confluent flasks with 10 ml 1× PBS per 225 cm^2 T-flask. To detach cells, add 5 ml PBS–1 mM EDTA per flask and incubate at 37 °C for 5–10 min or until cells round up and begin to detach.
2. Neutralize the PBS–EDTA solution quickly by adding growth medium (8 ml per 225 cm^2 T-flask) and collect the medium in a 15 ml falcon tube. Pellet the cells for 5 min at 282 × *g* at room temperature. Discard the supernatant. Resuspend the cells to form a homogenous suspension in 1× PBS.
3. For accurate plating, determine the cell concentration using a Nucleocounter and use 100 μl of the cell suspension to calculate viability using propidium iodide.
4. For normal cell line maintenance, split the cells 1:8 every 3–4 days when using 225 cm^2 T-flasks.
5. Seed the correct number of cells into a new culture flask, add 400 μg/ml of geneticin®, and place the culture flask in a CO_2 incubator, 5 % CO_2, 95 % humidity at 37 °C. For cell plating, use medium warmed to 37 °C at each step.
6. Lyse 50 μl of the cells ($<6 \times 10^6$/ml) by adding 50 μl lysis buffer (Buffer A). Add 50 μl of Buffer B to terminate cell lysis.
7. Dilute the cells to reach a density of 5×10^5 cells/ml with culture medium (*see* **Note 11**).
8. Plate 200 μl of cells (100,000 cells) per well into CellKey™ Standard 96 W Microplates.
9. Incubate for 10 min at room temperature without shaking or any kind of vibrations to let the cells settle, and then place in a 5 % CO_2 incubator overnight.

3.2.2 Day 2: Assay Buffer Preparation and Assay Plate Washing

1. Remove the cell plates from the incubator and remove the medium by "flicking" the plates (*see* **Note 12**).
2. Wash the cells twice with 100 μl of warmed CellKey™ buffer containing 1 % DMSO, and then add 135 μl CellKey™ buffer containing 1 % DMSO into each well (*see* **Note 13**).
3. Incubate for at least 45 min at 30 °C (*see* **Note 14**).

3.2.3 Compound Plate for Characterization of Antagonists or Inhibitors

Compounds have to be serially diluted 1:3 in 100 % DMSO to avoid any precipitation. Compounds should be prepared at ten times the final concentration to be tested.

1. Perform a serial dilution in the first 96-well polypropylene plate: To do this, initially dispense 15 μl of compound at 10 mM in 100 % DMSO into column 1. Then dispense 10 μl

of 100 % DMSO from columns 2–10 and transfer 5 μl of compounds from columns 1–2 and repeat the transfer sequentially through column 10. Then add 90 μl of CellKey™ buffer without DMSO into columns 1–10 and mix 2–3 times.

2. In the second polypropylene plate, dispense 180 μl of CellKey™ buffer without DMSO into columns 1–10 and transfer 20 μl of the previous solution (derived from **step 1**). Mix 4–5 times.
3. In the third polypropylene plate, dispense 40 μl per well from columns 1–10 of compound preparation from the 2nd polypropylene plate and dispense 40 μl per well of CellKey™ buffer in 1 % DMSO into columns 11 and 12.

3.2.4 Ligand Plate Preparation and Result Interpretation

Prepare the ligand solution at ten times the final concentration in assay buffer. We use the concentration corresponding to the EC_{80} (the concentration of ligand inducing 80 % of the maximal response) to determine the effect of antagonists (*see* **Note 15**).

1. Calculate the volume of ligand required: The volume per well = 40 μl. Therefore, at least 4 ml will be required for one 96-well plate (*see* **Note 16**). Example: If EC_{80} = 100 nM: Mix 50 μl of ligand at 100 μM in 5 ml of CellKey™ buffer in 1 % DMSO (*see* **Note 17**). Dispense 40 μl into all wells except in column 12 and dispense 40 μl of CellKey™ buffer in 1 % DMSO into column 12.
2. Include controls when characterizing inhibitor compounds. Positive controls correspond to the ligand which activates the receptor assessed. This measures the cell's ability to generate receptor-mediated responses through ligand activation. Generally 8 positive controls are included in the ligand plate in column 11. Assay buffer containing 1 % DMSO is used as the negative control for CellKey™ assays. Usually when characterizing antagonism, 8 negative controls (in column 12) are used in the plate to determine the basal cell responses. Other well-characterized inhibitors can be included in the list of compounds to be tested as reference compounds. Serial dilution of these known inhibitors should be prepared in parallel with tested compounds (the range of dilution should be adapted based on the known IC_{50}). These reference compounds may be available commercially such as the CCR2 antagonist sc-202525 from Santa Cruz or can be synthesized if the chemical structures are published, e.g., MK0812 from Merck & Co [8].
3. The response profile observed after ligand stimulation indicates the G-protein coupling for the studied receptor (*see* **Note 18**).

3.3 β-Arrestin

3.3.1 Agonist Dose–Response

1. Thaw the cells according to the manufacturer's instructions.
2. Plate 100 μl of cells at 8.33×10^5 cells per ml into each well of the provided 96-well tissue culture plate.

3. Then place the cells into a humidified incubator at 37 °C, containing 95 % air and 5 % CO_2 for 24–48 h prior to testing.
4. To determine the agonist EC_{80}: Perform a 11-point dose curve using at least duplicate wells for each dilution. Follow the guidelines as described in the instructions booklet. Briefly, prepare threefold serial dilutions of the agonist in the specific optimized cell culture (OCC) medium in the kit, containing the appropriate solvent (*see* **Note 1**). The concentration of each dilution should be prepared at 11× of the final screening concentration (i.e., 10 μl compound added to a volume of 100 μl). For each dilution, the final concentration of solvent should remain constant.
5. Transfer 10 μl from the serial dilution tubes to each well of the cells plated in the above section.
6. Incubate for 90 min at 37 °C.
7. Add 55 μl of prepared chemiluminescence detection reagent, according to the product booklet, per well, and incubate for 90 min at room temperature.
8. Read samples on any standard luminescence plate reader.
9. Use GraphPad Prism or other comparable software program to plot the agonist dose–response.
10. Determine the agonist's EC_{80} concentration from the curve. This concentration will be used in Subheading 3.3.2 to screen the antagonist compounds.

3.3.2 Antagonist Dose–Response

1. Thaw the cells according to the manufacturer's instructions.
2. Plate 100 μl of cells at 8.33×10^5 cells per ml into each well of the provided 96-well tissue culture plate.
3. Then place the cells in a humidified incubator maintained at 37 °C in 95 % air, 5 % CO_2 for 24–48 h prior to testing.
4. To test the antagonist compounds, dissolve the antagonist in the recommended vehicle or solvent of choice (DMSO, ethanol, PBS, water, or others) at the desired concentration.
5. Perform a ten-point dose curve using at least duplicate wells for each dilution by preparing threefold serial dilutions of antagonist compound in the specific OCC medium, containing the appropriate solvent. The concentration of each dilution should be prepared at 22× of the final screening concentration (i.e., 5 μl antagonist compounds added to make a final volume of 110 μl). For each dilution, the final concentration of solvent should remain constant.
6. Transfer 5 μl from serial dilution tubes to each well containing cells plated above (Subheading 3.3.2, **step 2**).
7. Incubate for 60 min at 37 °C.

8. Prepare the agonist compound in the specific OCC medium in the kit, containing the appropriate solvent (DMSO, ethanol, PBS, water, or others) at 22× of the EC_{80} concentration determined in Subheading 3.3.1 above (i.e., 5 μl agonist compound in a final volume of 110 μl). Add 5 μl of agonist compound to each well containing cells and antagonist.
9. Incubate for 90 min at 37 °C.
10. Add 55 μl of prepared chemiluminescence detection reagent, according to the product booklet, per well, and incubate for 90 min at room temperature.
11. Read samples on any standard luminescence plate reader.
12. Use GraphPad Prism or other comparable software program to plot the antagonist dose–response.

3.4 ELISA-Based Heparin Binding Assay

Carry out all procedures at room temperature unless specified otherwise.

3.4.1 Preparation of the Heparin Binding Plate

1. Dispense 200 μl of the heparin solution into each well of the heparin binding plate (*see* **Note 19**).
2. Seal the plate and wrap it in aluminum foil. Incubate the plate overnight.
3. The next day, prepare the blocking solution (0.2 % gelatin in PBS buffer): Prepare 30 ml blocking solution per plate. Add the gelatin into a glass beaker, and add PBS buffer to the final volume and mix with a magnetic stirrer. Incubate the solution in a 37 °C water bath until completely homogenous.
4. Discard the supernatant in the heparin binding plate by inverting the plate and add 200 μl SAB into each well. Repeat this washing procedure twice.
5. Repeat the washing procedure once more using PBS buffer. After the last wash, carefully tap the plate dry onto a tissue.
6. Add 250 μl blocking solution per well. Seal the plate and cover it in aluminum foil. Incubate the plate for 1 h at 37 °C (*see* **Note 20**).
7. Discard the supernatant, wash the plate three times with PBS buffer, and carefully tap the plate dry onto a tissue after the last wash.

3.4.2 CCL5 Heparin Binding Assay

1. Preparing the chemokine solution of interest, e.g., CCL5: Centrifuge the vial or Eppendorf tube containing the lyophilized chemokine prior to opening it. Add deionized distilled H_2O to obtain a final concentration between 1 and 10 mg/ml. Vortex and centrifuge briefly for 5 s using a bench mini centrifuge and prepare vials containing 10 μl aliquots of dissolved chemokine. Store vials at −20 or −80 °C.

2. Prepare a CCL5 dilution series in a 96-well round-bottom plate. Dilute CCL5 to a concentration of 1 μM in blocking solution (*see* **Note 21**) and prepare fourfold dilutions covering a dose range of 10^{-6} to 10^{-12} (*see* **Note 22**).
3. Transfer 100 μl of the dilution series in triplicate into the heparin binding plate wells. As a control, prepare wells containing blocking solution only without chemokine. Seal the plate and cover it with aluminum foil.
4. Incubate for 4 h at room temperature under gentle shaking (*see* **Note 23**).
5. Discard the supernatant by inverting the plate. Wash each well by adding 200 μl PBS buffer. Repeat the washing procedure twice and carefully tap the plate dry onto a tissue.
6. Dissolve the anti-human CCL5 antibody at 0.2 mg/ml in PBS buffer. Store the reconstituted antibody in appropriate aliquots at −20 °C.
7. Prepare the primary antibody solution by diluting the anti-human CCL5 antibody to 250 ng/ml in blocking solution (1:800 dilution). Transfer 200 μl of primary antibody solution into the heparin binding plate wells. Seal the plate, cover it in aluminum foil, and incubate it for 1 h at 37 °C.
8. Discard the supernatant and wash the plate three times with PBS buffer as described in **step 5**.
9. Prepare the secondary antibody solution by diluting the donkey anti-goat IgG-HRP antibody 1:10,000 in blocking solution. Add 200 μl of secondary antibody solution into the heparin binding plate wells. Seal the plate, cover it in aluminum foil, and incubate it for 1 h at 37 °C.
10. Discard the supernatant and wash the plate three times with PBS buffer as described in **step 5**. See Fig. 1 for the schema for chemokine–GAG binding prior to detection.

3.4.3 Detection of Heparin-Bound Chemokines

1. Add 100 μl HRP substrate TMB solution to each well and agitate slowly to mix. Incubate in the dark until the desired color intensity is reached (~20 min) (*see* **Note 24**).
2. Stop the reaction by adding 100 μl of 1 M sulfuric acid and measure the optical absorbance at 450 nm immediately using a fluorescent plate reader such as FLUOstar OPTIMA (*see* **Note 25**).
3. Analyze the data using the appropriate software, e.g., GraphPad Prism software. Data are expressed in optical absorbance at 450 nm ± standard error of the mean (SEM) on the *Y*-axis, and logarithm of CCL5 concentration in M on the *X*-axis (Fig. 3).

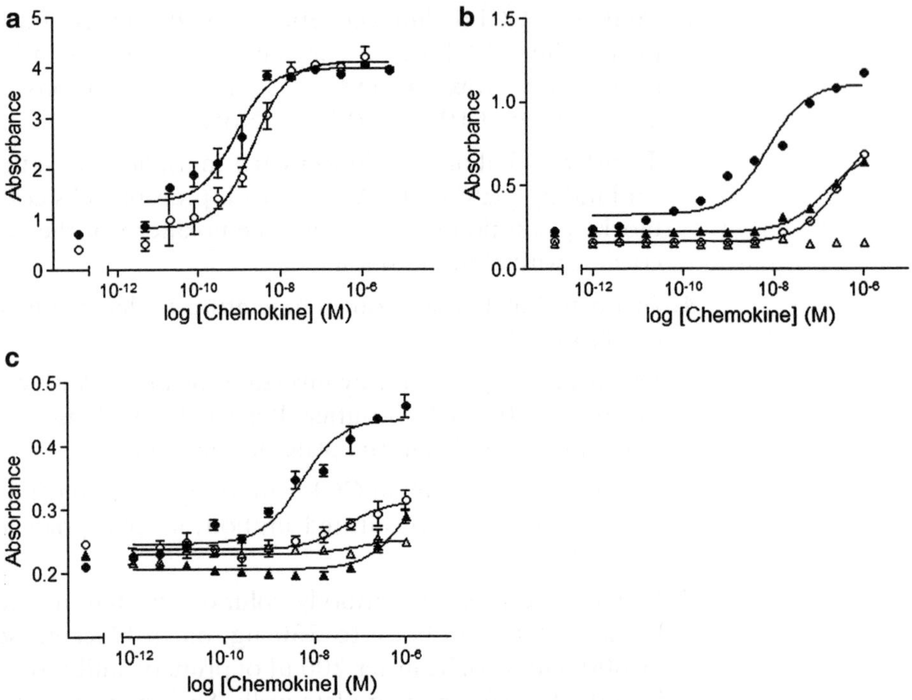

Fig. 3 Determination of specific CCL5 interaction with immobilized heparin. Comparison of the ability of CCL5 (*circles*) and the GAG-binding mutant 44AANA47-CCL5 (*triangles*) to bind to heparin-coated (*filled symbols*) and uncoated heparin binding plates (*open symbols*). Data are expressed in optical absorbance at 450 nm. (**a**) Chemokine and antibody in SAB. (**b**) Chemokine in blocking solution and antibodies in SAB. (**c**) Chemokine and antibody in blocking solution

3.5 Chemokine-Induced Erk Phosphorylation in Mouse, Human, and Cynomolgus Whole Blood

3.5.1 Whole Blood Collection and Activation

The assay described below uses murine CXCL13 to activate B cells in mouse blood.

1. Anesthetize mice in isoflurane chamber (3–5 %) and collect blood in heparinized tubes from retro-orbital sinus with a heparinized microcapillary tube.
2. Pool blood into a Falcon 15 ml polypropylene tube.
3. Aliquot 108 μl of pooled mouse blood into 5 ml Falcon polypropylene tubes, and then add 12 μl of 1× PBS or serial dilutions (1:3) of test inhibitors (*see* **Note 26**).
4. Incubate in a water bath at 37 °C for 5 min.
5. Prepare murine CXCL13 at a concentration of 5 μM in 1× PBS from the stock solution in water immediately prior to use.
6. Add 120 μl of different CXCL13 concentrations or PBS (as negative control) for 2.5 min at 37 °C to activate the target cells (*see* **Note 27**). Do not vortex or pipette up and down to prevent nonspecific activation.
7. Add 160 μl of 10 % formaldehyde in 1× PBS for 6 min on ice to stop the reaction by fixing the cells.

8. Add 2 ml of 0.16 % Triton X-100 (*see* **Note 28**) and incubate for 30 min at 37 °C to lyse red blood cells (*see* **Note 29**).
9. Centrifuge tubes at 460×*g* for 5 min at room temperature (RT) and wash once with 2 ml of 1× PBS.
10. After the wash, discard supernatant by inverting the tubes and transfer cell pellet in the residual volume of 1× PBS into a 96-well V-bottom plate (1 sample/well) for convenience.
11. Centrifuge plate for 4 min at 460×*g*, and wash once or twice with 200 μl PBS depending on whether red blood cells are observed.
12. Resuspend cells in 100 μl of 80 % methanol diluted in 1× PBS, seal, and store plate at −20 °C overnight.

3.5.2 FACS Staining

1. Add 120 μl of PBS to the plate and centrifuge for 4 min at 460×*g* at RT.
2. Wash cells with 200 μl of PBS per well, and then once more with PBS containing 4 % FCS.
3. Incubate at RT for 1 h with 50 μl of anti-phospho-Erk rabbit monoclonal antibody diluted 1:50 in 4 % FCS in 1× PBS per well.
4. Wash once with 200 μl 4 % FCS in 1× PBS per well by centrifugation (*see* **step 1**) and incubate cells at RT for 30 min with 50 μl of an antibody cocktail containing anti-mouse B220-PerCP-Cy5.5 diluted 1:50 and donkey-anti-rabbit IgG-Alexa 647 diluted 1:500 in the dark.
5. Wash the plate twice by adding 200 μl of 4 % FCS in 1× PBS and centrifuging as above.
6. Finally add 80 μl of 4 % FCS in 1× PBS and store the plate at 4 °C until ready to perform sample acquisition with a flow cytometer.

3.5.3 Flow Cytometry Acquisition

1. Acquire 10,000 B cell events per sample in the lymphocyte B220+ region, e.g., with the FACSCalibur™ flow cytometer using CellQuest Pro software.
2. Analyze flow cytometry data, e.g., using FlowJo software. Plot the phospho-Erk fluorescence histogram gated on B220+ cells.
3. Determine CXCL13-induced phosphorylated-Erk activity by measuring the percentage of B220+ phospho-Erk+ cells in each sample.
4. Generate excel tables with the percentage of phospho-Erk+ cells and export for IC_{50} calculation.
5. Calculate IC_{50} values using Excel fit software with 4 Parameter logistic model (model 205 under dose–response one site) by the following formula: $y = A + \{(B - A)/(1 + [C/x^{\wedge}D])\}$ where the *X*-axis represents the test inhibitor concentration in

non-logarithm format; the Y-axis represents the response (i.e., percentage of phosphorylated-Erk+ B cells); A is the bottom plateau of the curve, i.e., the final minimum y-value; B is the top of the plateau of the curve, i.e., the final maximum y-value; C is the x-value at the middle of the curve, which represents the log EC_{50} value when $A+B=100$; and D is the slope factor. In this model, a positive value is found when y decreases with increasing x and a negative value is found when y increases with increasing x.

3.6 Alexa-Chemokine Binding in Human, Mouse, Dog, and Cynomolgus Monkey Blood

1. Prepare a dilution series of a CCL2 antagonist in a 96-well round-bottom plate (*see* **Note 30**). The final concentrations should cover a dose range from 10^{-6} to 10^{-12} M.
2. Transfer 46 μl of blood into each well of the 96-well plate. We recommend performing each concentration tested in triplicate.
3. Add 2 μl of the serially diluted antagonist to the blood in each well or 2 μl of 2.5 % DMSO in 1× PBS 1× as a negative inhibition control (*see* **Note 31**).
4. Incubate for 30 min at RT.
5. Add 2 μl of 250 nM Alexa-CCL2 per well (resulting in a final concentration of 10 nM).
6. Incubate for 10 min at 37 °C.
7. Place the samples on ice and fix the cells with 125 μl of ice-cold fixative for 1 min.
8. Lyse red blood cells by adding 500 μl of ice-cold lysis solution and incubate the cells for 20 min on ice.
9. After complete red blood cell lysis, centrifuge the plate at 470 × g (1,500 rpm using the Heraeus Multifuge 1S-R centrifuge) for 5 min at RT.
10. Wash twice by adding 500 μl of 1× PBS into each well of the 2 ml 96-well plate, followed by a centrifugation at 470 × g for 5 min at RT.
11. Resuspend the cells in 100 μl of FACS buffer in each well.
12. Analyze the samples by flow cytometry for CCL2 binding (percentage of cells) in the monocyte and lymphocyte population, based on forward and side scatter characteristics (Fig. 4) (*see* **Note 32**). In the example of Fig. 5, the activity of the compound B is impaired by more than tenfold in blood compared to PBS.

3.7 Thioglycollate Peritoneal Recruitment with Receptor Coverage

1. The experimental design requires at least 6 mice/group. The sham group requires i.p. injection of PBS followed by treatment with the compound's vehicle. Control and treated groups are injected i.p. with thioglycollate and treated with vehicle (control) or the compound to be tested.

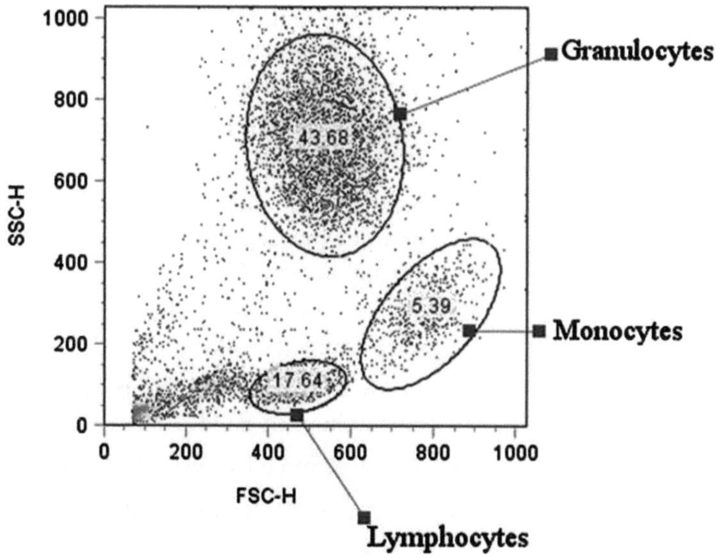

Fig. 4 Gating strategy to identify leukocyte populations. 100 μl of the cells resuspended in FACS buffer were run in a FACSCalibur machine. The analysis performed using FlowJo software allows a clear identification of the leukocyte population based on the size (*FSC-H* forward size scatter) and the granularity (*SSC-H* side size scatter)

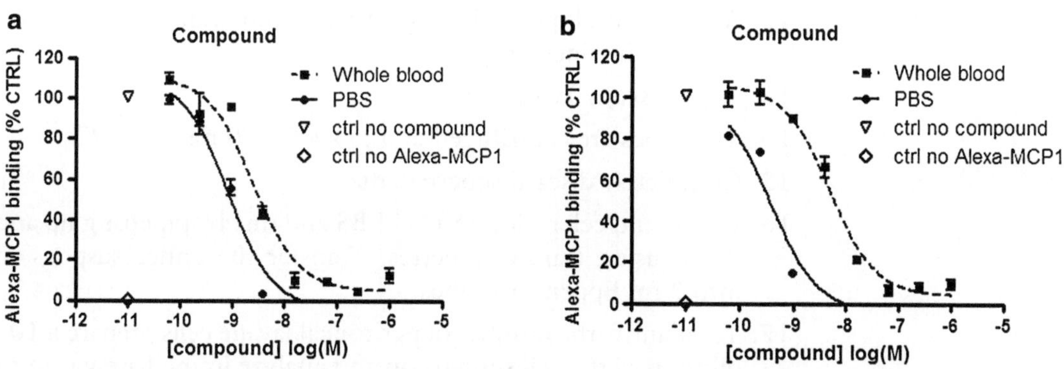

Fig. 5 Comparison of compound activity between whole blood and PBS. Compounds A and B were compared for their inhibition activity in PBS (*straight line, square symbol*) and human whole blood (*dash line, circle symbol*). Data is expressed in percent of total Alexa-CCL2 binding in the absence of any compound. As shown in the figure, while the activity of compound A is fivefold decreased in whole blood, the activity of the compound B is impaired by more than tenfold

2. Prepare compounds in appropriate vehicle (depending on the chemical properties of the compound and the route of administration).
3. Administer test compounds 30 min before thioglycollate injection and 24, 48, and 72 h after thioglycollate injection.

The frequency and route of administration, i.e., orally (p.o.) or i.p., will be dependent on the pharmacokinetic properties of the tested compound.

4. Weigh mice and inject 1 ml of 3 % thioglycollate or 1 ml of PBS (sham group) i.p.
5. Keep for a maximum of 96 h in Macrolon Type III cages (3 mice per cage) under the usual husbandry conditions of the animal facility.
6. To harvest peritoneal cells and collect blood, anesthetize mice with inhaled isoflurane (3–5 %).
7. Draw blood by cardiac puncture (under terminal anesthesia) and collect in EDTA-coated blood collection tubes.
8. Sacrifice mice by cervical dislocation.
9. Firmly affix mice unto a dissection plate and expose the abdominal skin.
10. Fill a 5 ml syringe with 5 ml cold PBS/EDTA. Stick the needle carefully into the peritoneal cavity in order to avoid damaging the peritoneal cavity contents. Inject the liquid into the abdominal cavity.
11. To improve the contact of the PBS/EDTA solution with the viscera, gently massage the abdomen with two fingers for a few seconds.
12. Aspirate the liquid with the syringe and transfer into falcon tubes maintained at 4 °C.
13. Repeat **steps 10–12** once.
14. The tubes are centrifuged at 425 × *g* for 10 min at 4 °C.
15. Completely discard supernatants.
16. Resuspend cell pellets in 1 ml PBS and mix by pipetting up and down using transfer pipettes. Transfer the entire suspension into 2 ml Eppendorf tubes.
17. To quantify the number of peritoneal lavage cells prepare a 10× dilution of the cell suspension and analyze immediately using a Beckman Coulter cell counter.
18. To determine receptor coverage, transfer 50 μl of the peritoneal lavage (in 1 ml PBS) or blood in triplicate per animal into the 96-well 2 ml plate.
19. Perform two controls per animal: Add 2 μl of the studied antagonist compound's saturating concentration (in 2.5 % DMSO) to 50 μl of blood or 2 μl of 2.5 % DMSO as the negative inhibition control (with a final concentration of 0.1 % DMSO).
20. Incubate for 30 min at RT.

21. Add 2 μl of 250 nM Alexa-CCL2 or Alexa-CCL4 (with a final concentration of 10 nM final) per well.
22. Incubate for 10 min at 37 °C to allow receptor binding.
23. Place the samples on ice and fix the cells with 125 μl of ice-cold light fixative for 1 min.
24. Lyse red blood cells by adding 500 μl of ice-cold lysis solution and incubate the cells for 20 min on ice.
25. After complete red blood cell lysis, centrifuge the plate at 470×*g* (e.g., 1,500 rpm using the Heraeus Multifuge 1S-R centrifuge) for 5 min at RT. Perform two washing steps with 500 μl of PBS, followed by a centrifugation at 470×*g* for 5 min.
26. Resuspend the cells in 100 μl of FACS buffer.
27. Analyze the samples by flow cytometry for CCL2 and CCL4 binding (percentage of cells) in the monocyte and lymphocyte population, based on forward and side scatter characteristics.

4 Notes

1. Chemokines are generally provided as lyophilized powders, usually from a trifluoroacetic acid solution. For this reason, always dissolve the chemokine in deionized distilled H_2O. They are very soluble proteins and can be easily dissolved at concentrations as high as 10 mg/ml, which is approximately 1 mM. Once in solution they can be diluted into the required buffer or medium. If the powders are dissolved directly in a buffer such as PBS, they may not go into solution, or only dissolve partially.
2. In this case, 10 μg per well of heparin sodium salt Sigma-Aldrich H3393, a mixture of polyanion chains with most chains in the range between 17,000 and 19,000 Da, was used. The assay can also be performed with other GAGs; however the appropriate concentration to saturate the plate's surface should be empirically determined (guided by the BD heparin binding plate supplier information).
3. 1 % weight/volume BSA in PBS can be used as an alternative blocking agent (kindly refer to the Iduron technical bulletin).
4. Sulfuric acid is highly caustic and needs to be handled with caution. Wear safety glasses, protect your skin, and work under a fume hood when handling it.
5. Cells were diluted 1:2 in trypan blue (e.g., Invitrogen) and living cells were counted using a cell counter such as the Countess® Automated Cell Counter (Invitrogen, http://products.invitrogen.com/ivgn/en/US/adirect/invitrogen?cmd=catProductDetail&productID=C10227).

6. The optimal cell concentration depends on the chemokine and the cells used. This should be determined prior to the assay.
7. Dilute to 1 μM in well A2 of the 96-well V-bottom plastic plate. Add 120 μl chemotaxis medium to wells A3–A11. Mix solution in the well A2 and transfer 60 μl (final 1:3 dilution) to the well A3. Repeat the transfer sequentially to the well A11. A1 and A12 would serve as controls containing the medium only.
8. The optimum migration duration depends on the chemokine and the cells used; this should be determined prior. It is usually between 30 min and 2 h.
9. SYBR green is a dye that binds to cellular DNA and allows an indirect but precise measurement of the number of migrated cells.
10. We have performed this automation using a Sciclone ALH 3000 workstation from Caliper Life Sciences, but alternative robotic workstations can be used.
11. The density of cells to be plated needs to be determined in a prior experiment during assay development.
12. "Flicking" plates will consist of removing the supernatant of each well by taking the plate in one hand and inverting it to throw the liquid away.
13. Be careful not to touch the center of the well which contains the gold electrode with the tips.
14. Buffer equilibration: Prior to any CellKey™ experiment, the culture medium needs to be replaced with assay buffer. This exchange could impact the basal impedance of seeded cells and an equilibration period of time should be applied at 30 °C to allow the cells to adapt to their new environment. This period ranges from 30 to 60 min depending on the cell line utilized. For CHO cell lines, this period is about 45 min, but do not hesitate to check that the cells are equilibrated before any experiment by recording the signal without any ligand addition.
15. The EC_{80} for the agonist should be determined in a prior experiment during assay development. A ligand dose–response has to be performed using the same experimental conditions.
16. When preparing the ligand plate, a 100 % response control can be added to check within the same plate whether the antagonist compounds are really effective, in addition to the EC_{80}.
17. The 100 times DMSO dilution in water can be omitted. As an alternative, the ligand can be diluted in buffer without DMSO, with the appropriate amount of DMSO added to make a 1 % concentration (the volume should be adjusted accordingly).

18. Analysis of CellKey™ response profiles: Before any analysis, the kinetic profile of the response must be assessed. This will give information on the G-coupling for the GPCR studied and distinguish a Gi-coupled from a Gq or Gs signature.
19. Nonspecific binding to the plate might differ among the chemokines used. To ensure accurate data, we strongly recommend preparing a control plate without immobilized heparin. This control plate should be used during the entire procedure under the same experimental conditions. We have observed nonspecific binding to the plate in the absence of heparin equivalent to the binding to heparin for certain chemokines, rendering this method unsuitable.
20. Depending on whether there is nonspecific chemokine binding to the plate, the percentage of gelatin in PBS as well as the duration of incubation can be modified to obtain optimal experimental conditions for each chemokine tested.
21. Gelatin can interfere with the chemokine binding to heparin or with antibody detection. However based on preliminary experiments, the chemokines and antibodies used in this assay were dissolved in blocking solution to minimize the observed nonspecific chemokine binding to the plate (Fig. 1). To study a decrease in the heparin–GAG interaction of CXCL11 mutants, no gelatin was added during the chemokine and antibody detection steps allowing a broader experimental range [9]. Thus the addition of a blocking agent is dependent on the experimental setup and the biological questions evaluated.
22. The conversion of the concentration in milligrams per milliliter (mg/ml) to molarity (moles per liter, M) is based on the formula concentration (g/l) ÷ molecular weight (grams per mole) = mole per liter.
23. The incubation period can be reduced depending on the chemokine used.
24. We recommend closely observing the color development as the incubation duration might be variable depending on the chemokine–antibody combination used. Once the wells that contain the highest concentration of chemokine solution added start to turn bright blue the reaction should be stopped. The addition of sulfuric acid turns the positive wells into a yellow color.
25. The optical absorbance should be read immediately as the signal intensity decreases with time. Precipitates may form, affecting the measurement's accuracy and sensitivity.
26. Do not mix blood too vigorously to avoid nonspecific activation.
27. Use a repetitive dispensing pipette to add the stimulating compound to optimize the precision associated with a short signal induction time.

28. Vortex tubes gently after the Triton-X-100 step to facilitate red blood cell lysis.
29. Ease fixation conditions (e.g., keep samples at 4 °C or reduce incubation time), if red blood cells do not lyse completely.
30. Place 20 μl of the CCR2 antagonist (1 mM) in well A1. Fill the wells A2–A12 with 15 μl of 100 % DMSO. Perform a 1:4 serial dilution by transferring 5 μl from wells A1 to A2. Mix well and repeat the dilutions sequentially to well A12. The line A contains now antagonists at 40 times the final concentration. Fill line B with 156 μl of PBS 1× and transfer 4 μl of the line A to the line B to obtain a 1:40 dilution required for the dose–response assessments.
31. This further 1:25 dilution in blood will result in final concentrations ranging from 10^{-6} to 10^{-12} M in 0.1 % DMSO.
32. The analysis of receptor coverage can be further detailed if markers of specific leukocyte subpopulations are used (e.g., CD3, CD19, Gr1, CD11b, CD11c, etc.).

References

1. Allegretti M, Cesta MC, Garin A, Proudfoot AE (2012) Current status of chemokine receptor inhibitors in development. Immunol Lett 145:68–78
2. Schall TJ, Proudfoot AE (2011) Overcoming hurdles in developing successful drugs targeting chemokine receptors. Nat Rev Immunol 11:355–363
3. Sachs D, Coelho FM, Costa VV, Lopes F, Pinho V, Amaral FA et al (2011) Cooperative role of tumour necrosis factor-alpha, interleukin-1beta and neutrophils in a novel behavioural model that concomitantly demonstrates articular inflammation and hypernociception in mice. Br J Pharmacol 162:72–83
4. Mortier A, Van DJ, Proost P (2008) Regulation of chemokine activity by posttranslational modification. Pharmacol Ther 120(2):197–217
5. Proudfoot AEI, Handel TM, Johnson Z, Lau EK, LiWang P, Clark L et al (2003) Glycosaminoglycan binding and oligomerization are essential for the in vivo activity of certain chemokines. Proc Natl Acad Sci USA 100:1885–1890
6. Mahoney DJ, Whittle JD, Milner CM, Clark SJ, Mulloy B, Buttle DJ et al (2004) A method for the non-covalent immobilization of heparin to surfaces. Anal Biochem 330:123–129
7. Proudfoot AE, Borlat F (2000) Purification of recombinant chemokines from E. coli. Methods Mol Biol 138:75–87
8. Xia M, Sui Z (2009) Recent developments in CCR2 antagonists. Expert Opin Ther Pat 19:295–303
9. Severin IC, Gaudry JP, Johnson Z, Kungl A, Jansma A, Gesslbauer B et al (2010) Characterization of the chemokine CXCL11-heparin interaction suggests two different affinities for glycosaminoglycans. J Biol Chem 285:17713–17724

Chapter 7

A Novel Approach to Quantify G-Protein-Coupled Receptor Dimerization Equilibrium Using Bioluminescence Resonance Energy Transfer

Irina Kufareva, Bryan Stephens, C. Taylor Gilliland, Beili Wu, Gustavo Fenalti, Damon Hamel, Raymond C. Stevens, Ruben Abagyan, and Tracy M. Handel

Abstract

Along with other resonance energy transfer techniques, bioluminescence resonance energy transfer (BRET) has emerged as an important method for demonstrating protein–protein interactions in cells. In the field of G-protein-coupled receptors, including chemokine receptors, BRET has been widely used to investigate homo- and heterodimerization, a feature of their interactions that is emerging as integral to function and regulation. While demonstrating the existence of dimers for a given receptor proved to be fairly straightforward, quantitative comparisons of different receptors or mutants are nontrivial because of inevitable variations in the expression of receptor constructs. The uncontrollable parameters of the cellular expression machinery make amounts of transfected DNA extremely poor predictors for the expression levels of BRET donor and acceptor receptor constructs, even in relative terms. In this chapter, we show that properly accounting for receptor expression levels is critical for quantitative interpretation of BRET data. We also provide a comprehensive account of expected responses in all types of BRET experiments and propose a framework for uniform and accurate quantitative treatment of these responses. The framework allows analysis of both homodimer and heterodimer BRET data. The important caveats and obstacles for quantitative treatment are outlined, and the utility of the approach is illustrated by its application to the homodimerization of wild-type (WT) and mutant forms of the chemokine receptor CXCR4.

Key words G-protein-coupled receptor, Chemokine receptor, CXCR4, Dimerization, Monomer–dimer equilibrium, Bioluminescence resonance energy transfer, BRET titration, BRET saturation

1 Introduction

Although it was originally assumed and corroborated by recent in vitro studies [1, 2] that some G-protein-coupled receptors (GPCRs) can function as monomers, there is now substantial evidence that many GPCRs homo- and heterodimerize. Further, it has been suggested that the dimer may be the minimal functional

Astrid E. Cardona and Eroboghene E. Ubogu (eds.), *Chemokines: Methods and Protocols*, Methods in Molecular Biology, vol. 1013, DOI 10.1007/978-1-62703-426-5_7, © Springer Science+Business Media New York 2013

unit [3–6]. Chemokine receptors, the focus of this volume, are no exception. One of the first clues that chemokine receptors oligomerize came from the discovery of a CCR5-Δ32 mutation [7]. CCR5 is one of the two main receptors involved in human immunodeficiency virus (HIV) entry into cells during the initial infectious phase of the disease, and it was found that individuals homozygous for the mutant were resistant to infection due to retention of the mutated receptor in the endoplasmic reticulum [8]. The fact that individuals heterozygous for CCR5-Δ32 also show delayed progression was then hypothesized to be caused by oligomerization of WT CCR5 with CCR5-Δ32, resulting in abnormal trafficking of the WT receptor to the cell surface. These data led to the notion that CCR5 might function as dimer at least in some contexts, which is now well established [8–10]. Similar phenotypic evidence for CXCR4 dimerization came from studies of the "warts, hypogammaglobulinemia, infections, and myelokathexis" (WHIM) syndrome which is an immunodeficiency caused by truncation of the receptor C-terminus that results in resistance to desensitization and internalization, and thereby enhanced signaling [11, 12]. Co-expression of WT CXCR4 with WHIM CXCR4 also leads to enhanced signaling and failure of the WT receptor to internalize upon stimulation with CXCL12, and this observation has been attributed to the ability of WT CXCR4 to dimerize with the WHIM variant [13, 14].

To date, many chemokine receptors have been shown to form homo- and heterodimers, not only with other chemokine receptors but also with GPCRs outside of the chemokine family [15, 16]. The functional consequences of these interactions have yet to be fully understood but include modulation of signaling responses such as transinhibition in ligand binding [17–20], as well as changes in G-protein coupling [10, 21]. Furthermore, the nature of the dimerization interfaces, the stability of the various oligomeric forms, the effects of the ligands on dimer equilibrium, conformation, and stability, and the diversity and plasticity of dimerization are even less well understood [22–30]. For example, all five of the crystal structures of CXCR4 complexed with a small-molecule antagonist or a cyclic peptide inhibitor revealed the same dimer interface involving helices V and VI [31]. Similarly, the recent structure of the μ-opioid receptor bound to an irreversible morphinan antagonist revealed a dimer stabilized by a four-helix bundle between helices V and VI [32], while the κ-opioid receptor bound to antagonist showed a dimer stabilized through helices I, II, and VIII [33]. Nevertheless, it is not clear whether these dimer interfaces are biologically relevant interfaces or artifacts of crystallization (Fig. 1), and thus, biochemical approaches are needed to complement structural studies [23, 30, 34–39]. Furthermore, higher order oligomers or array-like assemblies have been observed for some GPCRs in cryo-electron microscopy (EM) studies suggesting the existence of more than

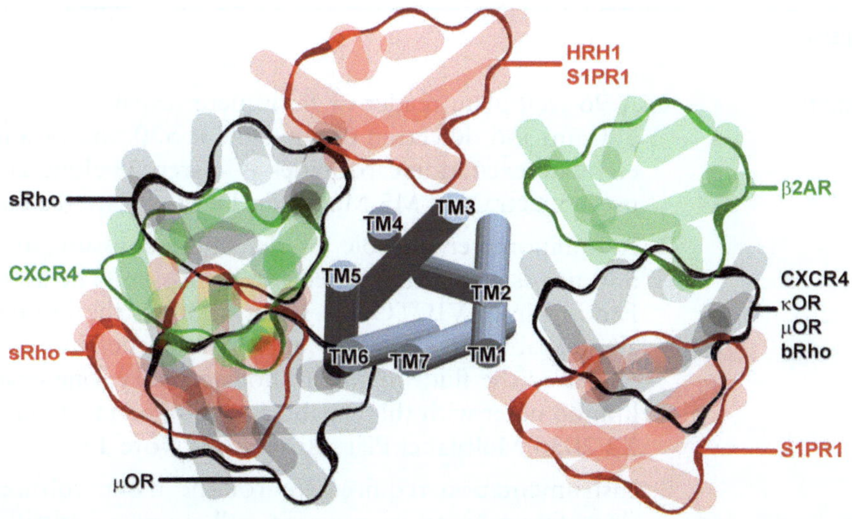

Fig. 1 Parallel GPCR dimer configurations observed by X-ray crystallography. The grey cylinders in the middle represent a superposition of GPCR monomers from multiple X-ray structures while the peripheral blobs illustrate the orientation of their crystallographic dimer partners. The structures are viewed from the extracellular side across the membrane. In different structures, the crystallographic dimer interfaces involve helices 1 and 2, 1 and 7, 4 and 5, 5 and 6, or 3 and 4. For some receptors like CXCR4, bRho, and S1PR1, more than one dimer configuration was observed. Several interface hypotheses are partially supported by biochemical studies [38, 39]. The methods described in this chapter allow a quantitative assessment of dimer interface mutants by BRET

one oligomerization interface on the surface of a particular GPCR. On the other hand, studies at physiological levels of receptor expression [27, 40] only convincingly corroborate the dimer, but not the higher oligomer theory.

In an attempt to address the relevance of the dimer observed in the crystal structures of CXCR4, we sought to introduce computationally designed mutations at strategic positions in the interfacial helices and to determine their effects on dimer stability using the bioluminescence energy transfer (BRET) assay. BRET is a biophysical method that has emerged as one of the main approaches for studying protein–protein interactions in cells [3, 41–49]. This process revealed some limitations in data quantification relevant to the analysis of all GPCRs, specifically when one wants to compare the relative dimerization propensity of different receptors or different mutants of the same receptor. Because BRET is a cell-based assay, the researcher has only a limited control of its input parameters (e.g., amounts of DNA for control of receptor expression), as the actual expression levels are inevitably affected by the cell machinery which in turn influences the assay readouts. Failure to take these effects into consideration may lead to misinterpretation of the assay results. We discuss this issue in more detail and propose an approach to quantitate expression levels. An analysis of both homodimer and heterodimer BRET data is presented.

2 Materials

2.1 Equipment

1. A 96-well plate reader or fluorimeter capable of excitation at 480 nm and detection of emission at 530 nm, or a luminometer with excitation capabilities as in **step 2** below. The authors use a SpectraMax M5 Molecular Devices plate reader.
2. A luminometer capable of detecting emission at 480 and 530 nm as well as unfiltered total emission. The authors use a Perkin Elmer VICTOR X Light 2030. If the luminometer is also capable of excitation at 480 nm, the fluorescent plate reader or the fluorimeter are not necessary; one example of a luminometer with this capability is the Perkin Elmer VICTOR X2 2030 Multilabel Plate Reader (*see* **Note 1**).
3. Instrumentation required for routine tissue culture research (biosafety cabinet, microscope, cell counter, centrifuge, water bath, etc.).

2.2 Reagents

1. Appropriate growth medium for the experimental cells to be used in BRET experiments.
2. Sterile 1 mL, 200 μL, and 20 μL micropipettes and sterile pipette tips for all three.
3. Any type of sterile 6-well plates commonly used for growing adherent cultured cells (*see* **Note 2**).
4. Sterile 0.5–1.5 mL DNA LoBind® tubes for preparation of transfection mixtures.
5. 1.5 mL tubes for cell containment after resuspension from the transfection plates and prior to plating in the BRET assay plate (*see* **Note 3**).
6. 96-well BioCoat/OptiLux plates (*see* **Note 4**).
7. A repeating pipette capable of dispensing 10 μL volumes, with tips (*see* **Note 5**).
8. Transfection reagent capable of efficiently transfecting the experimental cells for use in BRET experiments (*see* **Note 6**).
9. Transfection medium according to the transfection protocol; for example sterile serum-free Opti-Modified Essential Medium (Opti-MEM™) (Gibco).
10. Plasmid DNA encoding the proteins of interest fused to yellow fluorescent protein (YFP) and *Renilla reniformis* luciferase (Rluc), plasmid DNA encoding the fused Rluc–YFP construct, and empty vector plasmid DNA (*see* **Note 7**).
11. Phosphate-buffered saline (PBS): 137 mM NaCl, 2.7 mM KCl, 10 mM Na_2HPO_4, 2 mM KH_2PO_4, pH 7.4.
12. PBS with 1–5 mM EDTA for detachment of adherent cells.

13. BRET buffer: PBS with 0.1 % d-glucose.
14. Coelenterazine-h, 1 mM in 95 % ethanol. (This is the stock concentration: Generate diluted working solution just before use in the BRET experiment, *see* **Notes 8** and **9**).

2.3 Software

Data analysis software with nonlinear regression capabilities and graphing functionality, e.g., ICM (MolSoft), MatLab (MathWorks), or Prism (GraphPad).

3 Methods

3.1 General Description of the BRET Assay for Dimerization

BRET is an experimental technique designed to detect the spatial proximity of proteins in living cells. The proteins under question must be expressed in cells as fusion constructs with a bioluminescent donor protein and a fluorescent acceptor protein that is excited at the emission wavelength of the donor. In the process of substrate oxidation, the bioluminescent donor protein releases photons that are detected "as is" by the luminometer when the donor is in isolation. However, when appropriate acceptors are present in proximity (50–100 Å) to the donor protein molecules, the acceptors are excited into a higher energy state and subsequently emit the photons by fluorescence at a longer wavelength. One of the widely used donor/acceptor pairs is Rluc (maximum emission at 480 nm) and a genetic mutant of *Aequorea victoria* green fluorescent protein, YFP (excitation at 480 nm, emission at 530 nm) (Fig. 2). The experiment is initiated by adding a fixed amount of Rluc substrate (e.g., coelenterazine-h) to the cells co-expressing the donor/acceptor pair.

BRET has the advantage over fluorescence resonance energy transfer (FRET) methods because it eliminates artifacts associated with photobleaching and thus allows measurements in real time over an extended period relative to fluorescence resonance energy transfer [50]. Because it relies upon the addition of exogenous substrate for donor emission, problems associated with sample autofluorescence as well as direct excitation of the acceptor are minimized, leading to higher signal-to-noise ratios. Cell damage from the excitation light is also minimized [51]. Unlike fluorescent techniques involving chemical labeling, e.g., [52, 53], BRET is free of artifacts associated with labeling efficiency variations. However, BRET is prone to similar artifacts due to the presence of endogenous untagged receptor (*see* Subheading 3.4.5). Also, a bulky C-terminal fusion protein may affect interaction of the GPCR with downstream signaling partners, effectively altering its dynamics and membrane trafficking [54, 55] (*see* **Note 10**).

The energy transfer is quantified as the *ratio* of emission intensities at the acceptor emission wavelength (530 nm for YFP) and the donor emission wavelength (480 nm for Rluc). This ratio,

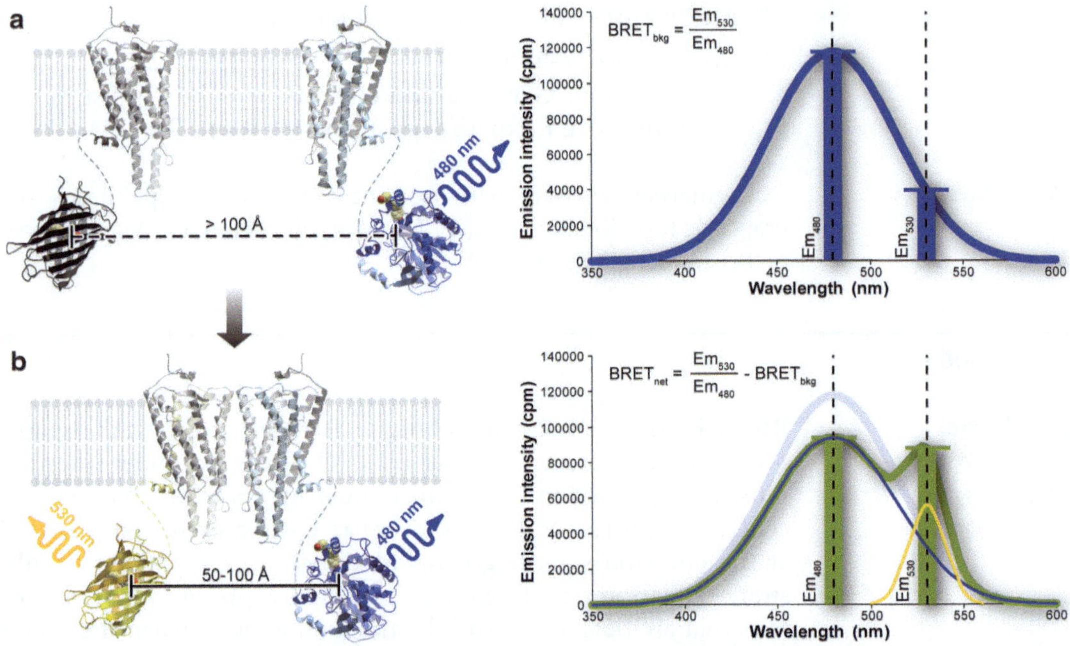

Fig. 2 The principle of the BRET experiment. (**a**) When the two proteins of interest do not dimerize, the luciferase emission spectrum is unchanged and the background ratio of emission intensities at 530 and 480 nm is observed. (**b**) When the proteins interact, the proximity of the Rluc and YFP molecules results in resonance energy transfer and a change in the shape of the emission spectrum with a reduction in the emission at 480 and an increase at 530 nm. The resulting combined emission spectrum is characterized by a larger ratio of emission intensities at 530 and 480 nm, with the degree of increase being indicative of the extent of energy transfer. The degree of transfer, in turn, is affected by several parameters discussed in the text

referred to as *BRET signal* or *BRET ratio*, is a simple numerical characteristic of the shape of the combined emission spectrum of the sample. When acceptors are absent, this ratio has a certain background value (~0.34, Fig. 2a); however acceptor molecules in the energy transfer proximity of the donors change the spectrum shape, shifting it towards longer wavelengths, thereby increasing the BRET ratio (Fig. 2b). Importantly, the shape of the combined emission spectrum appears independent of the total emission intensity (experimentally, we observe only a small variation in the BRET ratios across a wide range of luciferase emission intensities). Consequently, the BRET signal is relatively stable over time, sometimes for as long as 1 h, despite the fact that the absolute luminescence is subject to exponential decay as the substrate is used up.

Background BRET ($BRET_{bkg}$) is the BRET ratio characterizing Rluc emission when YFP is absent (*see* **Note 11**). Net BRET ($BRET_{net}$) is calculated as the difference between BRET and $BRET_{bkg}$. The maximal achievable value of BRET, $BRET_{max}$, is defined as $BRET_{net}$ when all available donor molecules are paired up with acceptor molecules. In the context of GPCR homodimerization, $BRET_{max}$ depends on the donor/acceptor distance and

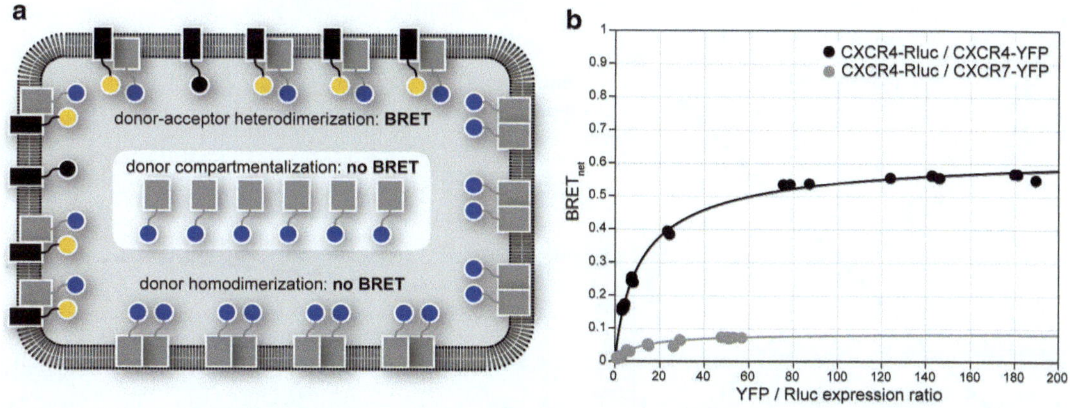

Fig. 3 Specifics of the BRET assay for heterodimerization studies. (**a**) A schematic illustration of some possible reasons for lower $BRET_{max}$ in a heterodimerization assay. Reasons include compartmentalization of the Rluc donor-tagged receptor so that it is sequestered from the YFP acceptor-tagged receptor and therefore unavailable for interaction, and a preference of the donor-tagged receptor for homodimerization, which may again render the donor-tagged receptor unavailable for interaction with the acceptor-tagged receptor. While acceptor homodimerization may also have detrimental impact on the BRET signal intensity, in practice it is addressed by simply increasing the acceptor expression levels. (**b**) BRET titration curves for the CXCR4/CXCR7 heterodimer compared to the CXCR4 homodimer. The much smaller curve for the heterodimer likely results from one or both of the reasons mentioned above

relative orientation within the dimer. The concept of $BRET_{max}$ intrinsically assumes that donors do not dimerize, which is not the case in GPCR dimerization studies; however, the relative fraction of donor–donor dimers and their contribution to the total BRET signal decrease with increasing acceptor/donor ratio.

In the context of heterodimerization, apart from the standard dependencies of signal intensity on donor–acceptor distance and orientation, $BRET_{max}$ is also affected by homo- vs. heterodimerization preferences of the donors as well as their cellular distribution and availability to form dimers with acceptors (Fig. 3); if donors preferentially dimerize with each other, or if they are present in some cellular compartments with no acceptors, their unchanged emission will substantially lower the overall observed BRET signal. Insufficient expression of the acceptors or a strong preference for acceptor homodimerization over heterodimerization with donors does not change the value of $BRET_{max}$ in theory; however, these issues may result in the inability of the system to *reach* $BRET_{max}$ in an experiment. In some cases, switching the donor/acceptor pair can help to clarify the true reasons behind the lowered $BRET_{max}$ in heterodimerization studies.

3.2 BRET Transfection Schemes and Titration Curves

One of the main problems with the BRET technique is that it measures spatial proximity of the molecules under study rather than their physical interaction. In experimental setups where the proteins are over-expressed in living cells, it is possible that nonzero $BRET_{net}$ results from simple crowding of the proteins on the cell surface,

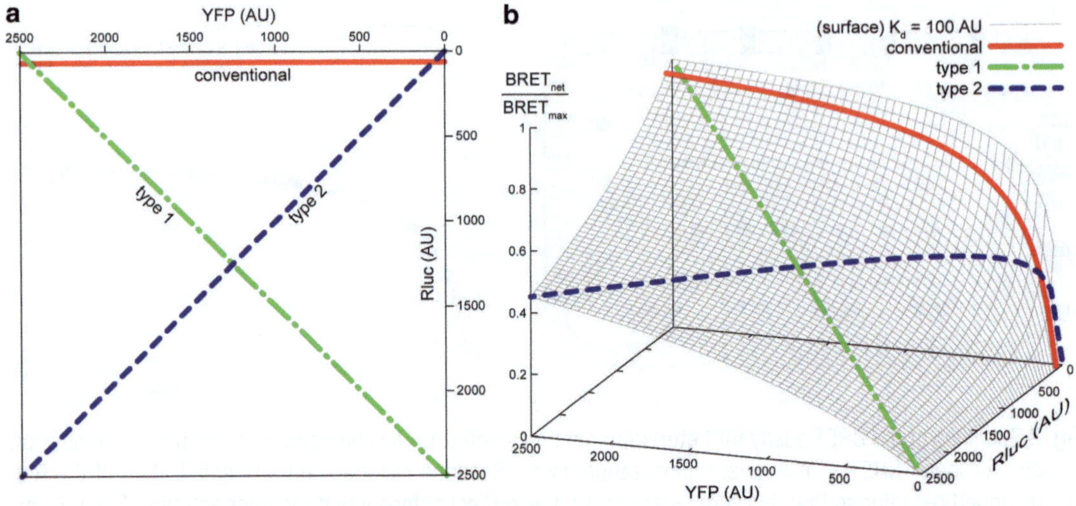

Fig. 4 Commonly used transfection schemes in BRET titration experiments. In a conventional scheme, the Rluc expression is held constant between the samples while YFP is varied. Type 1 and type 2 experiments require maintaining either total Rluc + YFP density (type 1) or YFP/Rluc ratio (type 2) constant while varying the other parameter (YFP/Rluc ratio for type 1 and total density for type 2). (**a**) A 2D representation of the assay schemes on the Rluc vs. YFP plane. (**b**) A projection of the three BRET transfection schemes onto a 3D BRET response surface. The surface illustrates how BRET response in a system with a given dimer equilibrium dissociation constant k_d depends simultaneously on the surface densities of Rluc- and YFP-tagged receptors. Notice that a typical BRET experiment following any of the three schemes corresponds to a 2D slice of the 3D BRET response surface

without direct interaction between them—the so-called bystander BRET. Consequently, nonzero BRET in a single sample cannot be unambiguously interpreted as evidence of protein dimerization. With a goal of distinguishing the nonspecific "bystander" effects from true dimerization, a rigorous BRET experiment involves a number of cell samples (from 7 to 12) transfected with varying amounts of donor- and acceptor-fused receptors followed by analysis of the resulting titration curves. Three schemes have been suggested for BRET saturation experiments [51, 56, 57] (Fig. 4a):

1. The "conventional" scheme involves sample transfection with a constant amount of donor-fused DNA and increasing amounts of acceptor-fused DNA in order to achieve variation in the acceptor/donor expression ratio. The DNA ratio is varied from 1 to as much as 250, where the higher ratio is required to fully saturate all donor molecules and obtain an accurate measurement of $BRET_{max}$. To avoid inter-sample variation due to transfection efficiency, it is important to complement the donor and acceptor DNA in all samples with empty vector for a constant total amount of DNA in each sample. Typical amounts of DNA are 0.01–0.1 μg of Rluc-tagged receptor DNA per well of a 6-well plate and up to 2.5–4 μg of YFP-tagged

construct. However, as noted below, the relative expression levels of each construct must be evaluated and taken into account since expression levels are not controllable.

2. In the so-called type 1 scheme, the total amount of donor + acceptor DNA is maintained constant between the samples, while the relative fraction of donors is varied. The advantage of this scheme is that, as long as the donor-tagged and acceptor-tagged receptors express equally well, any nonspecific bystander effects will be equally represented in all cell samples. The decision about the optimal total DNA amount in the type 1 scheme must be made based on the signal-to-noise ratio and the goals of the experiment. In some experiments, it is preferable to remain at physiological expression levels while other applications require overexpression.
3. In the "type 2" scheme, also sometimes described as "BRET dilution assay" [46], the acceptor/donor ratio is kept constant while varying the total amount of DNA. This allows one to study changes due to surface density increases. Optimal amounts of transfected DNA are calculated with a goal of sampling a wide range of expression levels: from as low as possible to get detectable expression to as high as the cells can handle without a significant reduction in viability.

In addition, a different approach to BRET titration involves competition between acceptor-tagged and untagged receptors (BRET competition experiments [43, 45, 46]), rather than saturation of donors with acceptors.

The specific BRET signal (i.e., resulting from dimer formation) is representative of two aspects of the system under study, namely: (1) the distance and relative orientation of donor and acceptor molecules when the dimer is formed, and (2) the fraction of donor molecules that are in dimers with acceptor molecules at the given surface densities of donors and acceptors. The experimental schemes outlined above are aimed towards controlling the latter aspect in a way that allows unambiguous analysis and result interpretation.

3.3 Typical BRET Assay

A detailed description of the BRET experimental protocol is provided [47–49]. In the present protocol we outline the assay with attention given to some caveats and useful intermediate readouts. To begin the experiment, plan a desired transfection scheme, compute the appropriate number of transfection samples, and determine positive and negative controls. Note that the protocol below is designed for measuring BRET with adherent cells that are in suspension during the assay. *See* **Note 12** for an alternative adherent format. BRET can also be performed on permeabilized cells, cell fractions, and partially purified receptors as described in [44].

3.3.1 Day 1: Seeding Cells

Seed cells in 6-well plates (*see* **Note 2** for discussion of alternative plates) in appropriate growth medium. Perform all steps under sterile conditions in a biosafety cabinet and use aseptic techniques. The optimal seeding density must be determined for each cell type and growth characteristics with the goal of obtaining 80–95 % confluence on the day of the assay.

3.3.2 Day 2: Transfection

Co-transfect cells with the appropriate Rluc-tagged and YFP-tagged receptor plasmids using an optimal cationic lipid-based transfection reagent (*see* **Notes 6**, **13**, and **14**). Perform all steps under sterile conditions in a biosafety cabinet and use aseptic techniques.

- For the conventional transfection scheme, transfect 7–12 wells as described above in Subheading 3.2. In addition, transfect a single well with Rluc-tagged receptor only (no YFP), using the same amount of Rluc DNA as with the assay samples to measure $BRET_{bkg}$. To do this, dilute the total amount of Rluc-fused receptor DNA that will be used for all of the wells into the volume of Opti-MEM™ medium recommended in Subheading 2.2. Add the varying amounts of YFP-fused receptor DNA to each tube one by one. Finally, add the appropriate amount of empty vector DNA required to each tube to obtain equal amounts of DNA for each well.
- For type 1 or type 2 transfection schemes, transfect 7–12 wells as described above in Subheading 3.2. In addition, include a matching well with the same amount of Rluc-tagged receptor only (no YFP) as in each Rluc/YFP sample for the determination of $BRET_{bkg}$. For type 1 BRET experiments, the Rluc- and YFP-fused receptor DNAs must be added individually to all samples. For the type 2 scheme, premix enough DNA for all wells to be transfected at the fixed YFP:Rluc ratio selected, and then divide up the DNA into tubes in the amounts intended for transfection. Complement each tube with empty vector DNA so that the total amount of DNA in each tube is equal.
- For all schemes, it is useful to include one or more wells transfected with an Rluc–YFP fusion construct [49] (*see* **Note 15**). These wells not only serve as positive controls for BRET but also provide means for the relative quantification of expressed Rluc-fused and YFP-fused receptors in the other samples as discussed below. 10 ng/well of the fusion DNA are typically used. However, a concentration series may also be valuable.
- For all schemes, to eliminate variability in transfection efficiency between the samples, use exactly the same total amount of DNA and transfection reagent for all wells, complementing it with empty vector if necessary.

- To ensure efficient *co-transfection* of plasmids, it is very important to combine and mix all DNA components of the transfection mixture *prior* to addition of the transfection reagent (*see* **Note 13**).
- Once all of the DNA mixtures are adequately mixed, add transfection reagent to the tubes according to the transfection reagent instructions. After adding the transfection reagent to the DNA, incubate at room temperature according to the reagent protocol. Finally add the transfection mixture to the wells in a drop-wise manner.

Depending on the cell type and transfection conditions, the maximum protein expression may occur at any time between 12 and 72 h post transfection; this time point should be chosen as day 3.

3.3.3 Day 3: Measurement of Cell Fluorescence and Luminescence

All steps may be performed on a laboratory bench. It is advisable to keep the cells and buffers on ice and to work as fast as possible to maximize cell viability.

1. Rinse cells attached to the plates with PBS once. This can be done by carefully pipetting 1 mL of PBS into each well of the 6-well plates, rocking the plates gently, and then aspirating the PBS.
2. Harvest cells in 1–2 mL of buffer and transfer to 1.5 mL tubes. Weakly adherent cells like HEK293T can be detached by gentle pipetting using BRET buffer (PBS + 0.1 % glucose). Other cell types require incubation with PBS + 1–5 mM EDTA at 37 °C for 5–10 min.
3. Wash twice with BRET buffer by centrifuging the cells at $300 \times g$ for 5 min, aspirating the supernatant, and resuspending in BRET buffer. Then count the cells and reconstitute all samples to 10^6 cells/mL; this number may need to be changed depending on the expression of the proteins under study (*see* **Note 16**).
4. Aliquot 90 μL of each sample and introduce into each well (in triplicate) of a 96-well plate. In order to avoid errors, it is important to leave empty wells between samples with high expected luminescence and those with low expected luminescence.
5. Measure total YFP fluorescence using excitation at 480 nm and emission at 530 nm for each well. The total fluorescence is used to quantify protein expression.
6. Add 10 μL of freshly prepared 50 μM coelenterazine-h in PBS to each well of the 96-well plate containing cells, moving across the entire plate as quickly as possible. Wait for 8–10 min to pass prior to analysis, bypassing the early expected rapid luminescence decrease (*see* **Notes 8** and **9**).
7. Using a PerkinElmer Victor X Light or comparable luminometer, measure the total luminescence with no filter selection for

0.5–1 s per well over the entire plate and then repeat three times. The goal of repeated measurements is to ensure that there is no significant luminescence decrease between the readings. With a relatively stable signal over time, the absolute luminescence values are considered comparable between the samples and used to quantify Rluc-tagged construct expression levels.

8. Repeat the emission measurements at 480 ± 15 and 530 ± 15 nm three to five times. This filtered luminescence measurement is necessary to calculate the BRET ratio.

3.3.4 Calculation of BRET Ratios

1. Calculate the BRET ratio for each well by dividing its emission at 530 nm by its emission at 480 nm.
2. Use the BRET ratio from YFP-free wells as $BRET_{bkg}$. If the experiment involves varying the amount of Rluc-tagged protein, plot $BRET_{bkg}$ against the total unfiltered luminescence in the YFP-free samples. If $BRET_{bkg}$ appears independent of the total luminescence, average the obtained $BRET_{bkg}$ values. If a consistent dose-dependent increase in $BRET_{bkg}$ is observed at low luminescence (which may happen with some luminometers), one may need to interpolate the data to obtain an accurate approximation of $BRET_{bkg}$ for each value of unfiltered luminescence. This issue is most relevant at low luminescence emission intensities in type 1 and type 2 experiments. Subtract the relevant $BRET_{bkg}$ from the BRET ratio in each YFP-containing well to obtain $BRET_{net}$.

Upon completion of the experiment, each well is characterized by the following values:

- DNA transfection amounts for Rluc- and YFP-tagged constructs.
- Total fluorescence (characterizes relative YFP-tagged construct expression).
- Unfiltered luminescence measured at a time point within the slow luminescence decay stage (characterizes Rluc-tagged construct expression).
- $BRET_{net}$.

Together, these values are utilized to quantify surface density-dependent GPCR dimer formation. Other parameters, such as number of cells per well, luciferase substrate concentration, and time of its addition, should be kept as constant as possible between the wells to eliminate unwanted variation.

3.4 Quantitative Interpretation of BRET Saturation Assay Results

In this section, we analyze the theoretical simulated shapes of titration curves in an attempt to gain a better understanding of the trends observed in each type of BRET experiment.

3.4.1 Conventional Framework for the Interpretation of BRET Saturation Data and Its Applicability Domain

In their 2006 publication in *Current Protocols in Neuroscience* [49], Hamdan et al. suggested a simple and accurate scheme for interpretation of curves resulting from a "conventional" BRET titration experiment (Fig. 5). According to this scheme, a specific interaction between the proteins labeled with donors and acceptors exhibits a saturable hyperbolic curve in coordinates of acceptor/donor ratio vs. $BRET_{net}$. From such a titration curve, it is possible to derive $BRET_{max}$ and the so-called $BRET_{50}$, i.e., the acceptor/donor ratio at which 50 % of the $BRET_{max}$ is achieved. Furthermore, when comparing two titration curves, variations in the $BRET_{max}$ suggest a conformational change between the binding partners while variations in $BRET_{50}$ are interpreted as changes in association affinity.

While this intuitive scheme has been instrumental for the interpretation of substantial BRET titration data, it is important to keep in mind that it is only applicable under conditions where:

- All curves are constructed with a single set of transfected samples and variation is created by changing the conditions at which the measurement is taken (e.g., agonist or antagonist addition).
- The samples express constant amount of donor protein and varying amounts of acceptor protein (i.e., the "conventional" scheme).

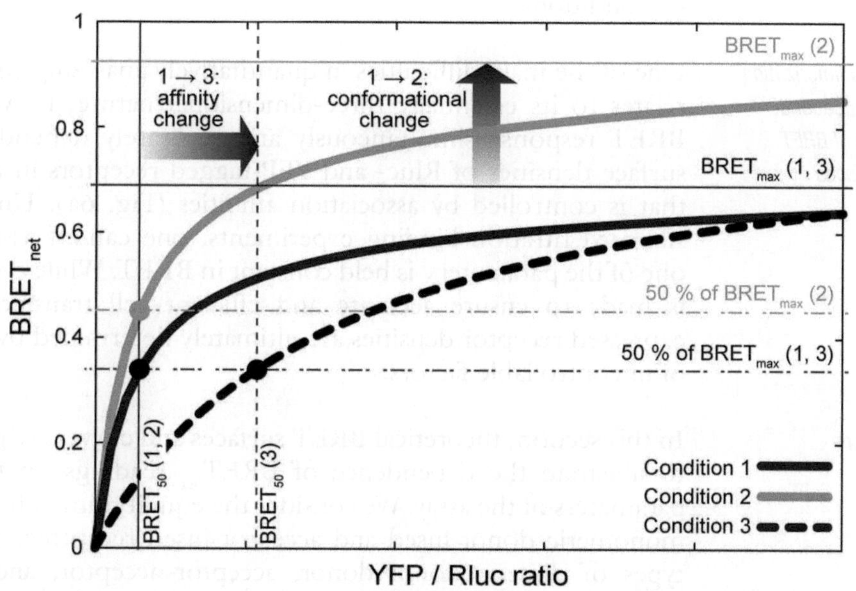

Fig. 5 Interpretation of curves obtained in a "conventional" BRET titration experiment as proposed in [49]. Variations in the $BRET_{max}$ suggest a conformational change between the binding partners while variations in $BRET_{50}$ are interpreted as changes in association affinity. The limits of applicability of this scheme are discussed in the text

- There is no competition between donors and acceptors for interaction with the same donors or the same acceptors (i.e., neither donor–donor dimerization nor acceptor–acceptor dimerization compete with the donor–acceptor complex formation).

Examples of experiments in which the above conditions hold include the monitoring of agonist-induced changes in the association of a GPCR with downstream signaling molecules, such as β-arrestins or G-proteins. In these applications, it is safe to assume that total amounts, cellular distribution, and luminescence/fluorescence properties of proteins under study do not change between the titration curves, and that donor–donor or acceptor–acceptor interactions do not prevent "productive" donor–acceptor pairs from forming.

These conditions, however, do not hold for experiments involving GPCR dimerization. With GPCR dimerization, there is considerable competition between nonproductive (donor–donor or acceptor–acceptor) pairs and productive donor–acceptor pairs. Furthermore, attempts to compare the curves obtained from different transfection samples, for example between mutants and wild-type proteins [30, 43, 44], are bound to fail because of inevitable variation in receptor surface densities.

Herein we seek to develop an alternative framework that allows quantitative interpretation of BRET data while avoiding the pitfalls associated with violation of the conventional framework applicability conditions.

3.4.2 Homodimerization Equilibrium Equations and Anatomy of BRET Titration Surfaces

One of the main difficulties in quantitatively analyzing BRET data relates to its essentially three-dimensional nature, by which the BRET response simultaneously and separately depends on the surface densities of Rluc- and YFP-tagged receptors in a manner that is controlled by association affinities (Fig. 6a). Unlike with standard titration binding experiments, one cannot assume that one of the parameters is held constant in BRET. While every effort is made to ensure accurate and efficient cell transfection, the expressed receptor densities are ultimately determined by a variety of uncontrollable factors.

Basic Equations

In this section, theoretical BRET surfaces and curves are generated to illustrate the dependence of $BRET_{net}$ readings on the input parameters of the assay. We consider the equilibrium of five species: monomeric donor-fused and acceptor-fused receptors, and three types of dimers: donor–donor, acceptor–acceptor, and donor–acceptor. The following notation is used:

- L and Y are total surface densities for Rluc-fused and YFP-fused receptors, respectively; $R = L + Y$ is the total surface density of the receptor.

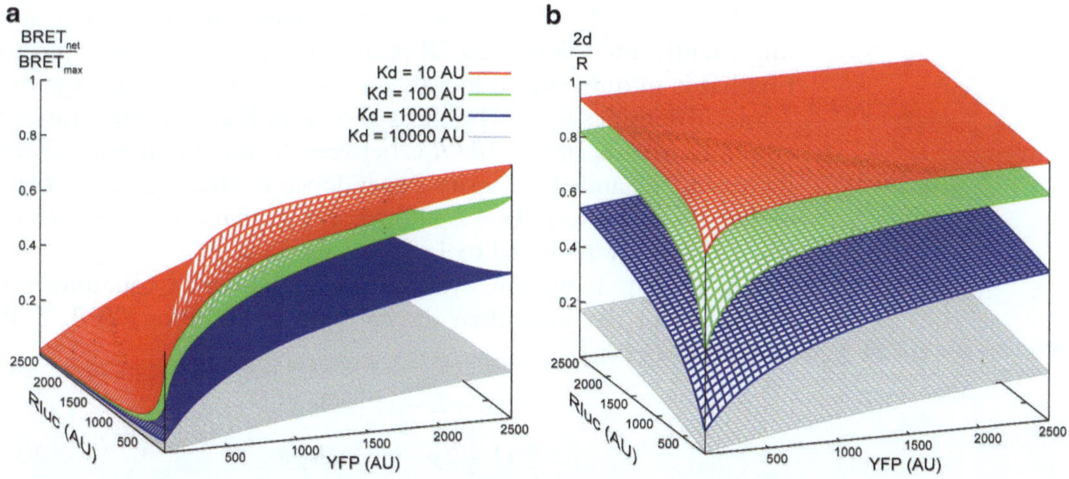

Fig. 6 Simulated surfaces of (**a**) BRET response and (**b**) fraction of receptor found in homodimers at varying surface densities of Rluc- and YFP-tagged receptors (shown on *horizontal plane axes*). BRET response (**a**) reflects the fraction of Rluc donor-tagged receptors that are in dimers with YFP acceptor-tagged receptors and does not take into account other possible homodimers (i.e., YFP/YFP or Rluc/Rluc homodimers). In contrast, fraction of receptor in dimers (**b**) represents all possible dimers. Varying association k_d results in changes of the surface shapes. These two types of 3D surfaces aid in understanding the pitfalls of quantitative interpretation of BRET data (see the text for details)

- l and y are surface densities of Rluc/Rluc dimers and YFP/YFP dimers; x is the density of Rluc/YFP dimers; the density of all dimers is given by $d = l + y + x$.
- $L-2l-x$ and $\Upsilon-2y-x$ are surface densities of monomeric Rluc-fused and YFP-fused receptors.
- k_d is a two-dimensional homodimer dissociation constant measured in the same units as receptor surface densities.

We assume that C-terminal fusions do not change the dimerization propensities of the proteins under study, and hence in this section, do not introduce separate dissociation constants for Rluc/Rluc, YFP/YFP, and Rluc/YFP dimers. However, because of combinatorial considerations, we need to multiply this universal k_d by a factor of 2 in all equations where the dimer components are symmetric [58].

The dynamic monomer–dimer equilibrium follows the standard equilibrium equation:

$$(R-2d)^2 = 2k_d d \tag{1}$$

This equation can be solved analytically to find the total dimer surface density, d, as a function of R and k_d:

$$d = \frac{2R + k_d - \sqrt{k_d^2 + 4Rk_d}}{4} \tag{2}$$

As expected, d increases with increasing R, but more importantly, the value of $2d/R$ representing the *fraction* of receptor that is in dimers also increases. $2d/R$ can be plotted against L and Y densities in a 3D plot as shown in Fig. 6b. For transient equilibrium association, $2d/R$ is between 0 and 1; it increases with increasing R, and the density-dependence of this increase is controlled by the dimerization k_d. For receptors that only exist in dimers, $2d/R$ is always equal to 1.

As for Eq. 1, we can write out the equilibrium equations for formation of the individual dimer types (Rluc/Rluc, YFP/YFP, and Rluc/YFP):

$$(L-2l-x)^2 = 2k_d l \tag{3a}$$

$$(Y-2y-x)^2 = 2k_d y \tag{3b}$$

$$(L-2l-x)(Y-2y-x) = k_d x \tag{3c}$$

By dividing Eqs. 3c by 3a and 3b by 3c, we obtain the following relations between monomeric and dimeric densities:

$$\frac{2y}{x} = \frac{Y-2y-x}{L-2l-x} = \frac{x}{2l} \tag{4}$$

The two sides of Eq. 4 can be manipulated separately to relate the surface densities of Rluc/Rluc and YFP/YFP dimers (l and y, respectively) to those of Rluc/YFP dimers (x):

$$l = x\frac{L}{2Y} \text{ and } y = x\frac{Y}{2L} \tag{5}$$

which can then be plugged into the mass equation, $d = l + y + x$, to yield

$$x = 2d\frac{LY}{(L+Y)^2} = 2d\frac{LY}{R^2} \tag{6}$$

Equations 2 and 6 link x, the density of donor–acceptor dimers, to the input parameters of the system, i.e., receptor densities L and Y and the homodimer dissociation constant k_d. Because only donor–acceptor dimers manifest themselves via resonance energy transfer and a change in the emission spectrum, and because these donor–acceptor dimers coexist in equilibrium with monomeric donor molecules and donor–donor dimers whose emission spectrum is unchanged, x is directly related to the experimental readouts of the BRET assay. The maximum achievable BRET ratio, $BRET_{max}$, occurs when the high surface density of the receptors effectively eliminates any monomeric species, and when the ratio of acceptors to donors is sufficiently high to eliminate donor–donor dimers, i.e., when $d = R/2 = (L + Y)/2$ and $x \approx L$. At any other condition, the ratio of $BRET_{net}/BRET_{max}$ is a reasonable approximation of x/L, i.e., of the fraction of donors that are engaged in dimers with acceptors. This relation is summarized in the simulated surface $BRET_{net}/BRET_{max}$ dependence on L and Υ densities for varying association affinities (Fig. 6a).

Expected Responses in Different Types of BRET Experiments

A single BRET experiment, conventional, type 1, or type 2, essentially traces a two-dimensional slice on the three-dimensional surface of association for a particular pair of constructs in question (Fig. 4b). Figure 7 provides a comprehensive account of expected responses in the three types of BRET transfection schemes in both monomer–dimer equilibrium and constitutive dimer situations. Several things should be noted:

- When a monomer–dimer equilibrium system is studied with a "conventional" experiment (Fig. 7a), the shape of the BRET response curve vs. YFP/Rluc ratio depends on the density of Rluc-tagged receptors. Higher expressors manifest in faster saturating curves in these coordinates.
- In contrast, if receptors only exist in dimers (Fig. 7b), Rluc density has virtually no effect on the shape of BRET response vs. YFP/Rluc ratio curve. This is because Rluc-tagged monomers do not exist and, consequently, do not contaminate the overall emission spectrum with unmodified luciferase emission.
- When a monomer–dimer equilibrium system is studied in a "type 1" experiment (Fig. 7c), the total receptor density (which is held constant between the samples) defines the shape and height of the curve of BRET response vs. YFP/Rluc ratio. It is important to note that while the type 1 curves appear to plateau at higher YFP/Rluc ratios, their asymptotic value is not $BRET_{max}$; instead, it simply characterizes the monomer–dimer equilibrium at the given surface density of the receptors. The type of response shown in Fig. 7c has been observed, for example, for β_2 adrenergic receptor homodimers in [51] (*see* Fig. 5 in the referenced paper).
- Studying a constitutive dimer in a type 1 experiment (Fig. 7d) produces a saturation curve in the YFP/Rluc ratio vs. BRET response coordinates that is independent of the total receptor density.
- When a monomer–dimer equilibrium system is studied in a "type 2" experiment (Fig. 7e), both the shape and the maximum value on the saturation curve in the YFP/Rluc ratio vs. BRET response plane depend on the selected ratio of YFP-tagged to Rluc-tagged receptor surface densities.
- With a constitutive dimer in a type 2 experiment (Fig. 7f), no saturation is observed, because the fraction of donors paired with acceptors is constant and defined by the (constant) YFP/Rluc ratio in the system.

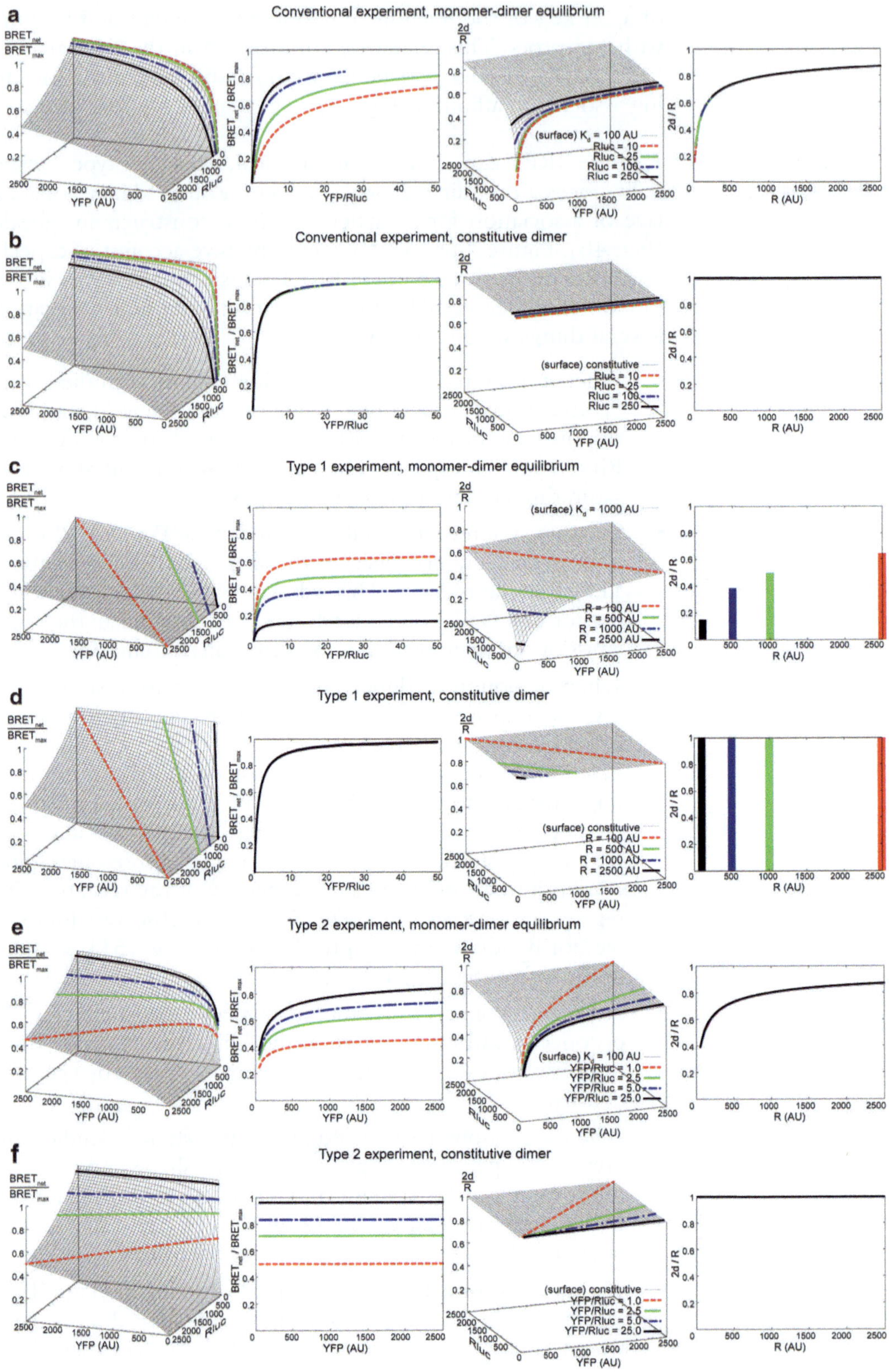
a
Conventional experiment, monomer-dimer equilibrium
BRET_net / BRET_max
YFP (AU)
Rluc
YFP/Rluc
2d / R
(surface) K_d = 100 AU
Rluc = 10
Rluc = 25
Rluc = 100
Rluc = 250
R (AU)
b
Conventional experiment, constitutive dimer
(surface) constitutive
Rluc = 10
Rluc = 25
Rluc = 100
Rluc = 250
c
Type 1 experiment, monomer-dimer equilibrium
(surface) K_d = 1000 AU
R = 100 AU
R = 500 AU
R = 1000 AU
R = 2500 AU
d
Type 1 experiment, constitutive dimer
(surface) constitutive
R = 100 AU
R = 500 AU
R = 1000 AU
R = 2500 AU
e
Type 2 experiment, monomer-dimer equilibrium
(surface) K_d = 100 AU
YFP/Rluc = 1.0
YFP/Rluc = 2.5
YFP/Rluc = 5.0
YFP/Rluc = 25.0
f
Type 2 experiment, constitutive dimer
(surface) constitutive
YFP/Rluc = 1.0
YFP/Rluc = 2.5
YFP/Rluc = 5.0
YFP/Rluc = 25.0

Potential BRET Quantification Errors

The three-dimensional interpretation of BRET response at the given surface densities of the Rluc- and YFP-tagged receptors highlights several possible reasons for errors in result quantification. For example, if for two construct pairs, the "conventional" experiment slices are taken at different donor surface densities, the resulting two-dimensional pictures may look different despite identical association affinities, or, on the contrary, look the same for pairs with different association affinities (Fig. 8a, b).

Moreover, donor and acceptor protein expression is often interdependent. For example, despite the fact that Rluc transfection levels are kept constant between the samples in the conventional transfection scheme, in homodimerization studies one can often observe a decrease in the unfiltered luminescence of the sample with increasing YFP expression (Fig. 8c). This may be due in part to competition between the two proteins for the same expression and folding machinery inside the cells: loading this machinery with overwhelming amounts of YFP-tagged receptor effectively takes the resources away from the Rluc-tagged receptor. Alternatively, in conventional scheme heterodimerization experiments, one may observe the opposite effect, i.e., an increase in luminescence with increasing YFP expression (Fig. 8d). A possible reason for this observation is a "chaperone" effect in which heterodimer formation makes the expression of Rluc-tagged receptor more favorable for a variety of reasons (stability, trafficking, etc.). Failure to relate the BRET signal to the studied protein expression rather than to the amount of transfected DNA may lead to serious misinterpretation of the results (Fig. 8e).

Interpretation of BRET Data in Terms of Fraction Receptors in Dimers

With simple rearrangements, Eq. 6 provides a way to tackle these issues by converting the observable BRET signal, $BRET_{net}/BRET_{max}$, into the standard scale of fraction of receptor found in dimers:

$$\frac{2d}{R} = \frac{BRET_{net}}{BRET_{max}} \times \frac{R}{Y} \tag{7}$$

The value of $2d/R$ plotted against R provides a reference frame for comparing association affinities between cases not comparable in the conventional scheme, e.g., different transfection samples,

Fig. 7 Simulated BRET responses for transient (**a**, **c**, **e**) and obligate (**b**, **d**, **f**) homodimers in a conventional (**a**, **b**), type I (**c**, **d**), and type 2 (**e**, **f**) BRET titration experiments. The 3D plots in the leftmost column illustrate BRET responses as a function of the surface density of Rluc-tagged and YFP-tagged receptors. Highlighted in thicker lines are slices taken at four Rluc densities for a conventional experiment (**a**, **b**), four total receptor densities for a type 1 experiment (**c**, **d**), and four YFP/Rluc ratios for a type 2 experiment (**e**, **f**). The second column contains projections of the same slices onto more traditional 2D coordinates. The third and fourth columns are results of translation of the corresponding BRET responses into monomer–dimer equilibrium schemes using the equations described in the text

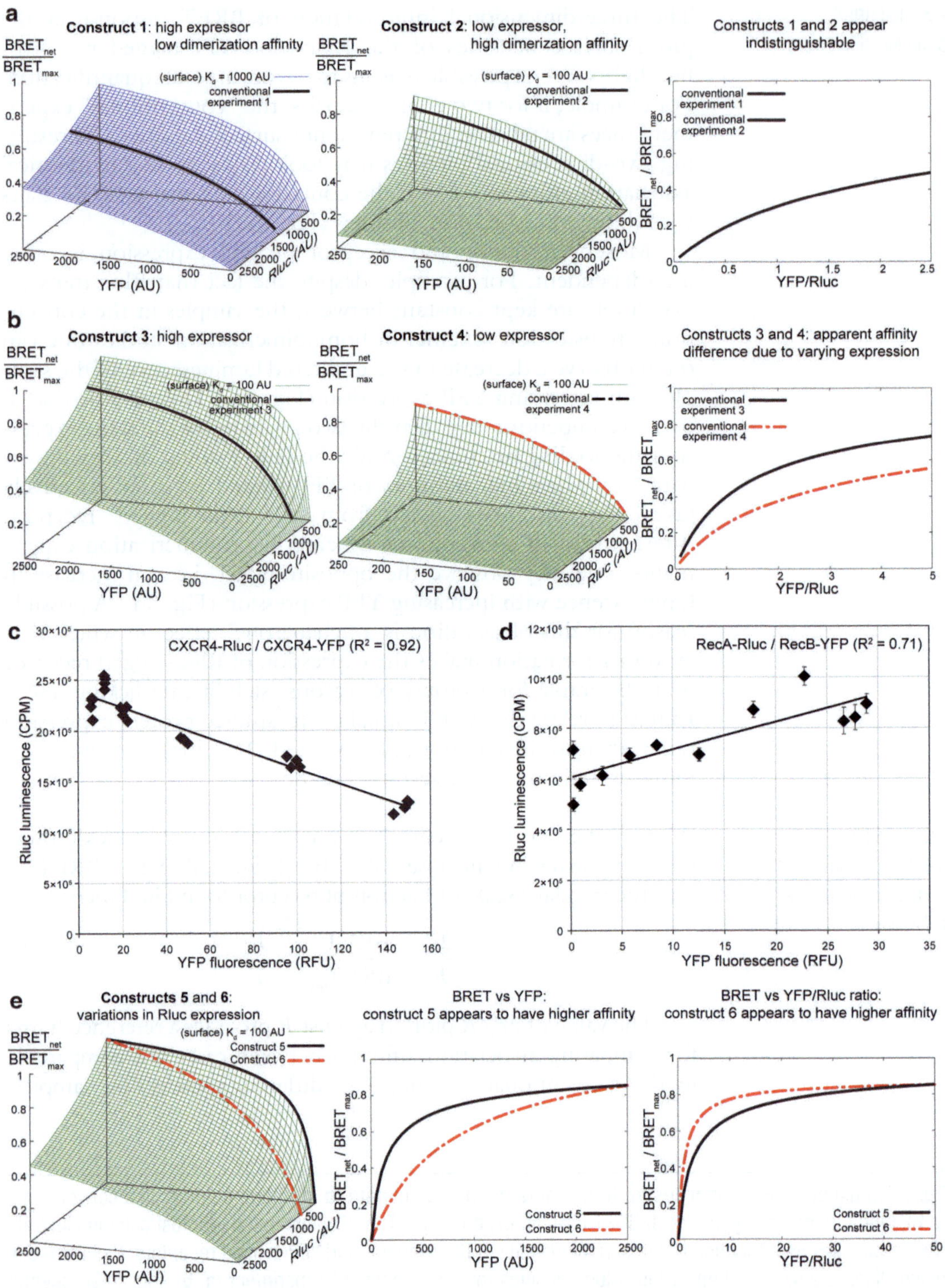

Fig. 8 Potential BRET quantification errors. (**a**) and (**b**): Disregarding variations in the expression levels between different constructs may lead to false conclusions about their relative homodimerization affinities. (**a**) Constructs 1 and 2 which have different homodimerization affinities and different overall expression levels can produce indistinguishable curves in coordinates of BRET response vs. YFP/Rluc ratio. (**b**) On the other hand,

including wild-type and mutant proteins. As Fig. 7 illustrates, regardless of the experiment type or the expression levels, application of Eq. 7 converts the data into the coordinates of $2d/R$ vs. R which unambiguously and reproducibly describe the dimerization equilibrium in the system (Fig. 7, columns 3 and 4). According to Eq. 7, to make this conversion possible, one only needs to know the ratio of surface densities of the acceptor-fused (Υ) and the total (R) receptor. Because $R = L + \Upsilon$, this question is equivalent to quantification of the relative densities Υ and L.

A Practical Way to Simultaneously Quantify Rluc- and YFP-Tagged Receptor Expression

The inevitable variation between Rluc-tagged and YFP-tagged construct expression makes it impossible to use the transfected DNA amounts as accurate approximations of construct expression levels. Relevant experimental readouts, i.e., sample luminescence and fluorescence, are often obtained on different instruments and are always measured in different units. However, the two quantities may be related to each other using the Rluc–YFP fusion construct and the BRET protocol readouts as follows:

- Use the experimental sample fluorescence for Υ (this would correspond to measurement of receptor expression in relative fluorescent units, RFU).
- Determine the fluorescence-to-luminescence normalization constant c as fluorescence (*fus*)/luminescence (*fus*) where *fus* is the experimental well with the Rluc–YFP fusion construct (*see* **Note 15**). The advantage of this construct is the guaranteed 1:1 stoichiometry between Rluc and YFP. It may be useful to have several wells transfected with varying amounts of the fusion construct and build a standard curve of luminescence vs. fluorescence.
- Quantify the density of the donor-fused receptor as $L = c \times$ Luminescence (unfiltered) in all wells (the density L of Rluc-tagged receptors is now expressed in RFU, similar to Υ).
- Quantify the total surface density as $R = L + \Upsilon$ (also expressed in RFU).

Fig. 8 (continued) constructs 3 and 4 with identical homodimerization affinities but different expression levels can translate into different curves on the BRET response vs. YFP/Rluc ratio plane. In both (**a**) and (**b**), the two constructs produce un-matching slices of their respective (distinct) 3D BRET response surfaces; this information is lost when the responses are plotted in 2D against YFP/Rluc ratio. (**c**) and (**d**): Despite the constant DNA transfection amounts, levels of Rluc-tagged receptor expression may decline (**c**) or increase (**d**) in samples with higher YFP-tagged receptor expression, as indicated by changes in the unfiltered luminescence. The plots represent data obtained in a CXCR4 homodimerization experiment (**c**) and in an undisclosed GPCR heterodimerization experiment (**d**). (**e**) Quantification errors due to disregarding the interdependence of expression levels of Rluc- and YFP-tagged constructs. If the Rluc-tagged receptor expression changes when the YFP-tagged receptor is varied, the problem can be thought of as taking nonparallel slices from the 3D BRET response surfaces, which can lead to artifacts of interpretation

When applying this scheme, all densities (R, L, and Y) will be expressed in RFU. Next, the data obtained using Eq. 7 can also be used to find k_d by fitting it into Eq. 1. This k_d will be expressed in the same units as R, L, and Y; in the scheme above it will be RFU. This k_d does not allow the absolute affinity quantification, but provides enough information for affinity comparison and construct ranking. However, if the densities are quantified by external methods in conventional units of cm^{-2} or mol/cm^2 [56], the dissociation constant will be expressed in these units as well.

Note that although using unfiltered luminescence for quantifying Rluc-tagged receptor expression levels is a standard practice [13, 56], it is not always entirely accurate. In reality, when luciferase molecules are in energy transfer proximity of the acceptors, a fraction of the emitted energy is absorbed by the acceptors and then emitted at a longer wavelength, contributing to the overall unfiltered emission of the sample. However, because the YFP quantum yield is less than one (it ranges from 0.6 to 0.8 for different variants [59, 60]), the resulting total unfiltered luminescence is lower than it would have been in the absence of acceptors in proximity. To the best of our knowledge, no studies have been conducted thus far to accurately quantify changes in unfiltered emission of Rluc as a result of energy transfer. One way to approach this question experimentally would be to compare the unfiltered emission of an Rluc–YFP fusion construct with that of a point mutation that renders YFP completely incapable of absorbing light in the Rluc emission spectrum.

The method of quantifying Rluc expression based on unfiltered emission relies on the assumption that the decrease in unfiltered emission due to energy transfer to the proximal YFP molecules is rather small. If the decrease is in fact significant, an additional correction factor is needed that relates the observed BRET ratio to the expected percentage decrease in unfiltered emission. Application of this factor will restore the observed unfiltered emission to its non-quenched value. The corrected emission can then be used for quantification of luciferase expression and further for BRET analysis as described above.

3.4.3 Example: Monomer–Dimer Equilibrium of the Chemokine Receptor CXCR4 and Its Point Mutants

To illustrate the utility of the framework described in Subheading 3.4.2, we applied it to study chemokine receptor CXCR4 homodimerization, as well as several single-point mutants. The mutants were computationally predicted to disrupt the dimer interface observed in the five crystal structures of the receptor discovered by Wu et al. [31]. Mutations were introduced in both YFP-tagged and Rluc-tagged receptor subcloned in pcDNA3.1 using the QuikChange site-directed mutagenesis kit (Stratagene). The constructs were transfected in HEK293T cells with TransIT LT1 (Mirus Bio) using conventional transfection protocols. Optimization of the transfection conditions led to robust and reproducible BRET signals for both WT and mutant receptors

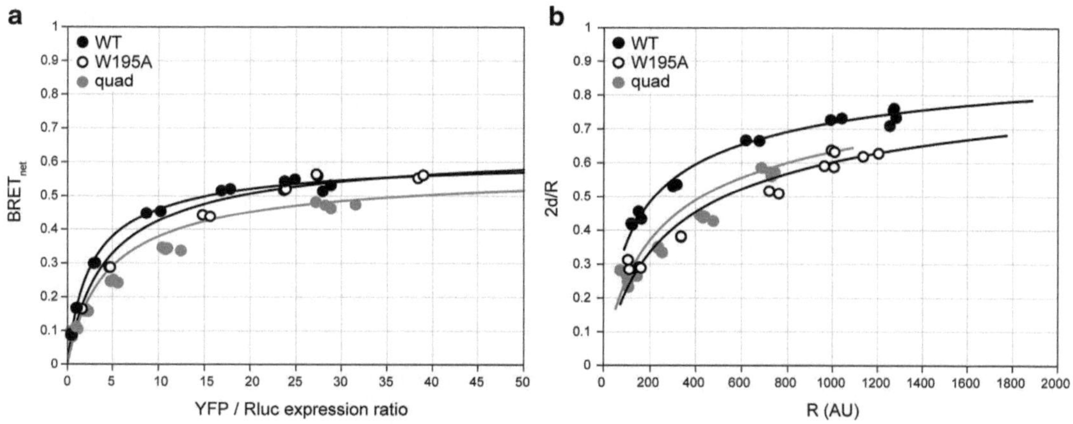

Fig. 9 Homodimerization of CXCR4 and selected point mutants of the TM5–TM6 interface observed in crystallography. (**a**) $BRET_{net}$ ratio is plotted against the YFP/Rluc expression ratio; (**b**) the same data converted to fraction of receptor in dimers plotted against the total surface density of the receptor. Differences between dimerization parameters have been clarified by correcting for variations in Rluc-tagged receptor expression between constructs and by converting the $BRET_{net}/BRET_{max}$ values into $2d/R$ values using Eq. 7 (see text). This example illustrates the advantages of the discussed additions to the data analysis. *quad = L194A/W195A/L267A/E268A

with $BRET_{max}$ exceeding 0.6 for the WT (Fig. 9a). A CXCR4-Rluc and $GABA_B$-YFP receptor pair was used as a negative control and provided a linear saturation response with $BRET_{net}$ never exceeding 0.025 (data not shown).

The CXCR4 mutations appeared to mildly affect both $BRET_{max}$ and $BRET_{50}$ (according to analysis by the conventional scheme); however these changes could not be translated into association affinity differences because of (1) variations in expression and (2) varying dependencies of luminescence (from the Rluc-tagged receptor expression signal) on YFP-tagged receptor expression between the mutant and WT constructs. By applying the equations described in Subheading 3.4.2, we were able to convert the BRET titration data into the common framework of dependence of receptor fraction in dimers ($2d/R$) on the total receptor density (R). This exercise clearly showed a difference in homodimerization affinities between WT and selected mutants (Fig. 9b). For example, about 2.5-fold higher surface density was required for the W195A mutant than for WT CXCR4 to achieve the same state on the monomer–dimer equilibrium scale. Further studies are required to elucidate whether additional mutations can be identified to further destabilize the dimer and the functional consequences of such mutations at physiological levels of receptor expression.

3.4.4 Quantifying Heterodimer Association Affinity

When studying heterodimerization, we can no longer assume that donor–donor, acceptor–acceptor, and donor–acceptor affinities are the same. To address this problem, we introduce numeric coefficients, a and b, to represent fold differences between donor–acceptor dimerization affinity and the other two affinities. The equilibrium equations for formation of the individual dimer types become

$$(L-2l-x)^2 = 2ak_d l \tag{8a}$$

$$(Y-2y-x)^2 = 2bk_d y \tag{8b}$$

$$(L-2l-x)(Y-2y-x) = k_d x \tag{8c}$$

By dividing Eqs. 8c by 8a and 8b by 8c, we obtain the following relationships between monomeric and dimeric densities:

$$\frac{2by}{x} = \frac{Y-2y-x}{L-2l-x} = \frac{x}{2al} \tag{9}$$

As seen with the homodimerization analysis, the two sides of Eq. 9 can be manipulated separately to relate the surface densities of Rluc/Rluc and YFP/YFP dimers (l and y, respectively) to those of Rluc/YFP dimers (x):

$$l = \frac{x}{2b}\frac{bL+x(1-b)}{aY+x(1-a)} \text{ and } y = \frac{x}{2a}\frac{aY+x(1-a)}{bL+x(1-b)} \tag{10}$$

Unfortunately, an equation analogous to Eq. 1 cannot be formulated for heterodimerization analysis because of dimer population heterogeneity. Nevertheless, the experimentally derived $BRET_{net}/BRET_{max}$ value remains a reasonable approximation of x/L. Therefore, the heterodimer association affinity can be deduced with the following three steps:

- Calculate the heterodimer surface density x from the BRET assay readouts as $x = L \times BRET_{net}/BRET_{max}$.
- Calculate the concurrent homodimer densities l and y using Eq. 10.
- Calculate the k_d from Eq. 8c.

Coefficients a and b must be optimized simultaneously with k_d for the best fit with the experimental data.

In the constitutive dimer system (i.e., when no monomer–dimer equilibrium is observed), the homodimer/heterodimer density ratios, l/x and y/x, are entirely dictated by the relative total surface densities of donors and acceptors. In other words, Eq. 5 holds for constitutive homodimers and constitutive heterodimers equally well.

3.4.5 Caveats and Limitations of Quantitative Analysis

Because the BRET assay is performed on living cells, it is prone to a variety of artifacts and result variability. In this section, we outline several major limitations in quantifying BRET titration curves.

- *Endogenously expressed untagged receptor*: If cells express the receptor of interest endogenously, the untagged receptor molecules may form dimers with Rluc-tagged or YFP-tagged receptors, thus complicating the analysis of the BRET titration data. Within the scope of BRET, it is impossible to discriminate Rluc-tagged monomers from their dimers with untagged receptor. Consequently, for quantitative BRET, it is preferable to choose a cell type with no to negligible background expression of the protein(s) under study.

- *Heterogeneity of protein expression*: The BRET response of each individual cell is determined by its receptor densities; however, the overall BRET signal is an average of signals coming from all the cells in the assay well. If the cells express varying amounts of the target proteins, the cumulative signal will represent an average of widely distributed signals from individual cells effectively lowering the signal-to-noise ratio. It is therefore important to choose a system (cell type and transfection method) that guarantees homogenous expression of proteins under study. To evaluate expression homogeneity, one can transfect a single sample with YFP-tagged receptor (maximal amount suggested by the transfection protocol) and test cells after 24 or 48 h for YFP expression by flow cytometry. Untransfected cells should be used as a control. A narrow peak that is separated from the control (untransfected) peak by at least 1.5–2 logarithmic units usually corresponds to efficient homogenous transfection and expression, while multiple peaks or a wide distribution of fluorescence intensities suggest that either the transfection protocol requires optimization or the cell type is inappropriate for quantitative BRET.
- *Cellular distribution of dimer partners*: The inherent limitation of BRET technology is its inability to distinguish signals coming from the cell surface from those originating from intracellular compartments. This is especially important when one attempts to quantify heterodimer association affinity. For example, if the Rluc-tagged dimer partner is selectively retained within the endoplasmic reticulum, either inherently or as a result of the C-terminal fusion, its non-modified emission will significantly affect the overall $BRET_{net}$ signal in the system, both in terms of $BRET_{max}$ and apparent association affinity. However, unequivocal discrimination of this effect from high-affinity donor–donor association or from purely a consequence of heterodimer asymmetry may require at least switching the fusion partners and ultimately other experimental approaches such as fluorescence microscopy, radioligand binding, etc.
- *Non-dimer-related BRET*: The three transfection schemes described in Subheading 3.2 only partially address the question of specificity of the dimer interactions; they provide means to distinguishing dimer "bystander" BRET resulting from receptor overexpression and consequent crowding. However,

they do not eliminate possibilities of forming higher order oligomers or receptor clustering as a result of partitioning into membrane microdomains. Total internal reflection fluorescence (TIRF) microscopy [40], luciferase- and YFP-complementation studies [61], and other techniques must be used to ultimately resolve these uncertainties [62, 63].

- *Estimation of Rluc expression levels*: This is important to obtain accurate dimer density quantification and k_d. As described in Subheading 3.4.2, while YFP-tagged receptor expression estimation is straightforward, determining Rluc-tagged receptor expression involves unit conversion that, in cases of high BRET, may be nontrivial. Improved methods to directly determine Rluc-tagged receptor expression levels are therefore desirable.

3.5 Application of the Quantitative Framework for Analysis of Dimerization Mutants

Structural biology studies of GPCRs have significantly increased since 2007 with the emergence of many new structures in various states of activation. Some of these such as CXCR4, the μ-opioid receptor, and the κ-opioid receptor have revealed parallel dimers in their asymmetric units or in the crystallographic cells [31–33]. Characterization of the functional relevance of these dimer interfaces via site-directed mutagenesis requires methods for quantitative assessment of relative dimer stability. Our method for analytical interpretation of BRET data should facilitate such comparisons by taking into account expression levels of the BRET receptor pairs.

4 Notes

1. The choice of luminometer depends on the specifics of the assay that will be employed. The author's laboratory utilizes a VICTOR X Light 2030 luminometer (Perkin Elmer) with a 2-channel liquid injection system for ligand addition. This instrument lacks fluorescent excitation capabilities; therefore, a separate plate reader with fluorescence reading capability is used to quantify energy acceptor expression. A liquid injection system is useful for the BRET^2 assay, which recommends immediate data collection following substrate addition. The time the instrument requires to begin the assay and move from well to well may affect data collection for temporally sensitive protein–protein interactions. Lastly, temperature control options and overall instrument sensitivity should be taken into account when deciding upon a plate reader. Alternative instruments commonly used for BRET measurements include the Mithras LB 940 (Berthold Technologies) and the LUMIstar Omega (BMG Labtech) plate readers.
2. In some cases, 12- or 24-well plates may be used, rather than 6-well plates, to seed and transfect cells for use in the BRET

experiments. The plate selection should be based on the requirement to provide adequate cell numbers for the planned experiment.

3. The size of the tubes used in the assay for harvesting and washing depends on the volume in which the cells are resuspended from the transfection plates, the cell numbers expected from each plate well, and the requirements of the centrifuge that is used to spin down the cells when switching buffers and normalizing cell concentrations.
4. The optimal plate type for BRET assays depends on the luminescence- and fluorescence-detecting equipment used. For example, we use a fluorescent plate reader which requires clear bottom plates and a luminometer that requires non-clear bottom plates. Therefore, the authors use BioCoat/OptiLux clear bottom plates and cover the bottom with white adhesive tape between fluorescent plate reader and luminometer.
5. For dispensing coelenterazine-h into 96-well plate wells, we have found it useful to use a repeating pipette for this purpose, as manually pipetting the substrate into the wells one by one is too slow given the rate of coelenterazine-h oxidation by luciferase.
6. We have had success with TransIT LT1 (Mirus Bio) in HEK293T cells and with Lipofectamine 2000 (Invitrogen) and TransIT CHO kit (Mirus Bio) in CHO cells for transfection.
7. We use a working DNA concentration between 50 and 500 ng/μL depending on the specifics of the assay. Furthermore, improved transfection efficiencies have been observed when fresh DNA working solutions are regularly prepared from higher DNA concentration stocks. We have routinely used both the Marligen Biosciences PowerPrep HP Plasmid Maxiprep Kit and the Promega PureYield Plasmid Midiprep System to generate high-concentration DNA stocks.
8. Coelenterazine-h is very sensitive to decomposition. The substrate should be aliquoted into appropriate volumes to prevent repeated exposure to light and water vapor in the air. For long-term storage we have found it useful to store stock concentrations (1 mM) of coelenterazine-h at –20 °C, in a desiccant-filled container.
9. Improvements to the BRET assay have been made that increase signal sensitivity, reduce background signal, and extend the reliable lifetime of the measurement (see [48] and references therein). The Rluc/YFP energy donor–acceptor pair utilized in the method reported herein represents the first-generation ($BRET^1$) development of the assay. The second-generation BRET ($BRET^2$) assay utilizes a modified coelenterazine

(DeepBlueC or coelenterazine-400a) substrate for Rluc which emits at ~420 nm as well as a mutant GFP (GFP[2] or GFP10) as the energy acceptor which emits at ~510 nm. This second-generation substrate/construct combination results in greater spectral separation between Rluc donor emission and YFP acceptor emission, thereby reducing background signal. However, it also results in a lower overall signal and shorter time frame for reliable data collection because of substrate instability. To address the substantially weaker signal, a series of mutations have been introduced into Rluc (Rluc2 and Rluc8) that significantly increase signal intensity. For longer kinetic measurements, a chemically protected coelenterazine substrate (EnduRen) has been developed. It requires cellular esterases to convert the substrate prior to oxidation. EnduRen enables measurement of the BRET ratio for up to 24 h; however, it requires performing the assay at 37 °C.

10. You need to ensure that labeling of the receptor's C-terminus does not affect its function. In our experience, addition of both small and large amino-acid motifs or proteins to the C-terminal tail can affect proper receptor trafficking and function. The inertness of the donor–acceptor labels on receptor function needs to be confirmed by ligand binding and functional assays.

11. Although referred to as $BRET_{bkg}$, this measurement is not actually BRET at all as it characterizes the situation with no energy transfer: it is the ratio of the luminescence emission intensities of Rluc at 530 and 480 nm in the absence of acceptors.

12. The assay described above is intended for cells in suspension (including dissociated adherent cells in suspension). To assay adherent cells, follow the same transfection protocol as described above except 6 to 24 h following transfection, lift, count, and split $0.5–2 \times 10^6$ cells in appropriate cell growth medium into a 96-well tissue culture-treated assay plate and incubate overnight. The following day, gently wash the cells without disturbing them from the bottom of the plate and replace medium with 90 μL/well BRET buffer. Carry out subsequent steps as detailed in the protocol.

13. At the transfection step, it is critically important that all cells receive the same total amount of DNA (*transfection efficiency*), as well as the same amounts of Rluc-tagged and YFP-tagged DNA (*co-transfection efficiency*). With liposome-forming cationic lipid reagents in particular, make sure that all DNA components of the transfection mixture are thoroughly mixed together *prior* to adding the transfection reagent. This increases the liposomal particle homogeneity upon reagent addition.

14. We developed a set of schemes based on serial dilutions to produce the multiple transfection mixtures for BRET experiments. They are especially useful when dealing simultaneously with a large number of samples. The schemes require that all working DNA stock concentrations are 50 ng/μL. The stocks include Rluc-tagged receptor plasmid (referred to as simply Rluc in this Note), YFP-tagged receptor (a.k.a. YFP), and empty vector (we use pcDNA3.1). The schemes are designed to deliver a total of 2.5 μg of DNA into a single well of a 6-well plate. The transfection mixtures can be prepared in 0.5–1.5 mL DNA lo-bind tubes or in wells of a 96-well plate (see below).

    Transfection scheme for a conventional BRET experiment. This scheme produces 7 samples per construct, 6 of them being for data points (in tubes 1–6) and one being a YFP-free sample for $BRET_{bkg}$ determination.

    - Aliquot 83.33 μL of YFP in tube 6, and 50 μL of pcDNA in tubes 1–3 and 7.
    - Take 33.33 μL from tube 6, pipet it into tube 3, and mix thoroughly.
    - Take 33.33 μL from tube 3, pipet it into tube 2, and mix thoroughly.
    - Take 33.33 μL from tube 2, pipet it into tube 1, and mix thoroughly.
    - Take 22.22 μL from tube 1, pipet it into tube 4 (empty), add 27.78 μL YFP, and mix thoroughly.
    - Take 11.11 μL from tube 1, pipet it into tube 5 (empty), add 38.89 μL YFP, and mix thoroughly.

    Result: 50 μL DNA at 50 ng/μL (e.g., a total of 2.5 μg DNA) in each of the tubes 1–7; pure pcDNA in tube 7, pure YFP in tube 6, and decreasing YFP/pcDNA ratio in all other tubes. The obtained ratios optimally sample the range for construction of a conventional BRET response curve.

    - To each tube, add the desired (constant and small, no more than 5 μL) volume of Rluc, and mix thoroughly. Adding 5 μL of the 50 μg/μL Rluc stock will result in a maximum YFP/Rluc ratio of 10 (in tube 7). It is usually preferable to sample higher ratios. To do so, decrease the added volume of Rluc. To reduce small-volume pipetting errors dilute the 50 μg/μL working Rluc stock tenfold or lower and pipet a proportionally larger volume.

    Transfection scheme for a type 1 BRET experiment. This scheme results in 6 samples per construct (in tubes 1–6) and 6 samples with identical amounts of Rluc but no YFP for $BRET_{bkg}$ determination (tubes 1*–6*). There will be a total of 2.5 μg receptor DNA per sample.

- Aliquot 83.33 μL of Rluc in each of the tubes 1 and 1*, 50 μL of YFP in tubes 2–6, and 50 μL of pcDNA in tubes 2*–6*.
- Take 33.33 μL from each of the tubes 1 and 1*, pipet it into tubes 2 and 2*, respectively, and mix thoroughly.
- Take 33.33 μL from each of the tubes 2 and 2*, pipet it into tubes 3 and 3*, respectively, and mix thoroughly.
- Take 33.33 μL from each of the tubes 3 and 3*, pipet it into tubes 4 and 4*, respectively, and mix thoroughly.
- Take 33.33 μL from each of the tubes 4 and 4*, pipet it into tubes 5 and 5*, respectively, and mix thoroughly.
- Take 33.33 μL from each of the tubes 5 and 5*, pipet it into tubes 6 and 6*, respectively, and mix thoroughly.
- Discard 33.33 μL from tubes 6 and 6*.

Result: 50 μL DNA in each of the tubes 1–6 with pure Rluc in tube 1 and increasing YFP/Rluc ratios in tubes 2–6. The ratios range from 0 to almost 100 using this scheme, with more attention given to the lower part of the curve. Additionally, there will be pure Rluc in tube 1* and increasing pcDNA/Rluc ratios in tubes 2*–6* (these samples will be used for $BRET_{bkg}$ calculations at various luminescence intensities).

Transfection scheme for a type 2 BRET experiment. This scheme produces 6 samples per construct (in tubes 1 through 6) and 6 samples with identical amounts of Rluc but no YFP for $BRET_{bkg}$ determination.

- Mix YFP and Rluc DNA to the desired proportion for the minimal total volume of 150 μL in a separate tube (make a little extra to account for possible pipetting errors). Repeat in another tube, using pcDNA (instead of YFP) and Rluc.
- Aliquot 83.33 μL of Rluc–YFP mixture into tube 6. Aliquot 83.33 μL of Rluc–pcDNA mixture into tube 6*. Aliquot 50 μL of pcDNA in each of the tubes 1–3 and 1*–3*.
- Take 33.33 μL from each of the tubes 6 and 6*, pipet it into tubes 3 and 3*, respectively, and mix thoroughly.
- Take 33.33 μL from each of the tubes 3 and 3*, pipet it into tubes 2 and 2*, respectively, and mix thoroughly.
- Take 33.33 μL from each of the tubes 2 and 2*, pipet it into tubes 1 and 1*, respectively, and mix thoroughly.
- Take 22.22 μL from each of the tubes 1 and 1*, pipet it into tubes 4 and 4*, respectively, add 27.78 μL Rluc–YFP mixture to tube 4 and 27.78 μL Rluc–pcDNA mixture to tube 4*, and mix thoroughly.

- Take 11.11 μL from each of the tubes 1 and 1*, pipet it into tubes 5 and 5*, respectively, add 38.89 μL Rluc–YFP mixture to tube 5 and 38.89 μL of Rluc–pcDNA mixture to tube 5*, and mix thoroughly.

Result: 50 μL DNA in each of the tubes 1–6; increasing amounts of receptor DNA from 1 to 6 with the constant YFP/Rluc ratio. Matching amounts of Rluc DNA in tubes 1* to 6* (these samples will be used for $BRET_{bkg}$ calculations at various luminescence intensities).

As an alternative to tubes, DNA mixtures can be prepared in a sterile untreated round- or conical bottom 96-well plate using a 50 μL multichannel pipette. Following DNA mixture preparation, use a multichannel pipette to add transfection medium and reagents as described in the reagent protocol. In most 96-well plates, a single well can hold up to 300 μL. In our protocol, the final transaction mixture volume for a single well in a 6-well plate does not exceed 250 μL; therefore, a 96-well plate is appropriate.

15. The Rluc–YFP fusion construct was produced in our lab by fusing the YFP coding sequence into the Rluc vector (generously provided by the Bouvier lab) downstream of the Rluc coding sequence.
16. Increasing the number of cells in each well of a 96-well assay plate may improve the total luminescence/fluorescence yield from the sample, and consequently reduce signal-to-noise ratio, without affecting the BRET ratio. For example, if one or both of the proteins of interest express poorly, more cells may be required to provide a sufficient signal for luminometer detection. Alternatively, if the signal saturates the detector then fewer cells should be used. Trial and error may be required to optimize assay cell numbers.
17. In all types of BRET assays, it is preferable to test a wide range of acceptor expression levels. Moreover, reliable measurement of $BRET_{max}$ requires the Rluc-tagged receptor population to be saturated with YFP. A protein that expresses poorly will yield small variability in expression and result in a small YFP/Rluc expression ratio. Therefore, the lower expressing receptor should be fused to Rluc while the better expresser should be fused to YFP.
18. Depending on the particular receptor interaction being studied, it may be advantageous to conduct the BRET assay at room temperature (as described above) or at 37 °C and to pre-equilibrate the cells to that temperature prior to substrate addition. At room temperature, the fluidity of the cellular membrane, lateral receptor mobility, and the function of potentially important scaffolding proteins may be affected compared

to more physiological temperatures. When measuring constitutive dimerization, the assay temperature may or may not affect the apparent affinity; however, when monitoring the effect of ligand on receptor dimerization, the receptor's ability to diffuse within the membrane may become significant. Adjusting the assay temperature will shorten the time frame in which reliable measurements can be made in the $BRET^1$ and $BRET^2$ assays due to more rapid coelenterazine-h substrate oxidation. On the other hand, prolonged incubation with substrate and data collection at 37 °C is necessary when using the EnduRen substrate (*see* **Note 9**).

19. The effect of ligand modulation on GPCR dimerization remains unclear and thus far appears highly receptor or ligand dependent, or dependent on both. However, the BRET assay can be a useful tool in determining whether a ligand induces a change in dimer affinity or a conformational change between a preformed dimer complex. As long as the same population of transfected cells is used, the "conventional" quantification scheme [49] can be applied to address this issue because the only difference is whether ligand is present or not; all other parameters are the same.
20. Negative controls: Another receptor that does not dimerize with the receptor of interest can be useful to distinguish between specific and bystander BRET. Many class C receptors, e.g., $GABA_B$, form obligate homodimers and have low to no affinity for class A receptors [49, 56].

Acknowledgements

This work was supported by the National Institute of Health PSI:Biology grants U01 GM094612 (T.M.H. and R.A.) and U54 GM094618 (R.C.S.) and by NIH grants R01 GM081763 (T.M.H.) and R01 GM071872 (R.A.). C.T.G and B.S. were supported by the Cellular and Molecular Pharmacology Training Grant T32 GM007752. D.H. was supported by NIH NRSA grant F32 GM083463. B.W. was supported by NIH grant R01 AI100604 and the grant 11JC1414800 awarded by Science and Technology Commission of Shanghai Municipality. The authors would like to thank Jeffrey Velasquez (TSRI) for help with molecular biology, Tam Trinh and Kirk Allin (TSRI) for help with the baculovirus expression, and M. Bouvier (University of Montreal) for the Rluc and YFP coding sequence containing vectors used to produce all of our BRET constructs. We also thank Pascale Charest (University of Arizona) for valuable discussions regarding the BRET assays and Goran Pljevaljcic (TSRI) for constructive comments on the manuscript.

References

1. Whorton MR, Bokoch MP, Rasmussen SG, Huang B, Zare RN, Kobilka B, Sunahara RK (2007) A monomeric G-protein-coupled receptor isolated in a high-density lipoprotein particle efficiently activates its G protein. Proc Natl Acad Sci USA 104:7682–7687
2. Bayburt TH, Leitz AJ, Xie G, Oprian DD, Sligar SG (2007) Transducin activation by nanoscale lipid bilayers containing one and two rhodopsins. J Biol Chem 282: 14875–14881
3. Angers S, Salahpour A, Bouvier M (2002) Dimerization: an emerging concept for G-protein-coupled receptor ontogeny and function. Annu Rev Pharmacol Toxicol 42:409–435
4. Springael JY, Urizar E, Parmentier M (2005) Dimerization of chemokine receptors and its functional consequences. Cytokine Growth Factor Rev 16:611–623
5. Terrillon S, Bouvier M (2004) Roles of G-protein-coupled receptor dimerization. EMBO Rep 5:30–34
6. Milligan G (2004) G-protein-coupled receptor dimerization: function and ligand pharmacology. Mol Pharmacol 66:1–7
7. Liu R, Paxton WA, Choe S, Ceradini D, Martin SR, Horuk R, MacDonald ME, Stuhlmann H, Koup RA, Landau NR (1996) Homozygous defect in HIV-1 coreceptor accounts for resistance of some multiply-exposed individuals to HIV-1 infection. Cell 86:367–377
8. Benkirane M, Jin DY, Chun RF, Koup RA, Jeang KT (1997) Mechanism of transdominant inhibition of CCR5-mediated HIV-1 infection by ccr5delta32. J Biol Chem 272:30603–30606
9. Issafras H, Angers S, Bulenger S, Blanpain C, Parmentier M, Labbe-Jullie C, Bouvier M, Marullo S (2002) Constitutive agonist-independent CCR5 oligomerization and antibody-mediated clustering occurring at physiological levels of receptors. J Biol Chem 277: 34666–34673
10. Mellado M, Rodriguez-Frade JM, Vila-Coro AJ, Fernandez S, Martin de Ana A, Jones DR, Toran JL, Martinez AC (2001) Chemokine receptor homo- or heterodimerization activates distinct signaling pathways. EMBO J 20:2497–2507
11. Hernandez PA, Gorlin RJ, Lukens JN, Taniuchi S, Bohinjec J, Francois F, Klotman ME, Diaz GA (2003) Mutations in the chemokine receptor gene CXCR4 are associated with WHIM syndrome, a combined immunodeficiency disease. Nat Genet 34:70–74
12. Lagane B, Chow KY, Balabanian K, Levoye A, Harriague J, Planchenault T, Baleux F, Gunera-Saad N, Arenzana-Seisdedos F, Bachelerie F (2008) CXCR4 dimerization and beta-arrestin-mediated signaling account for the enhanced chemotaxis to CXCL12 in WHIM syndrome. Blood 112:34–44
13. Percherancier Y, Berchiche YA, Slight I, Volkmer-Engert R, Tamamura H, Fujii N, Bouvier M, Heveker N (2005) Bioluminescence resonance energy transfer reveals ligand-induced conformational changes in CXCR4 homo- and heterodimers. J Biol Chem 280:9895–9903
14. Balabanian K, Lagane B, Pablos JL, Laurent L, Planchenault T, Verola O, Lebbe C, Kerob D, Dupuy A, Hermine O, Nicolas JF, Latger-Cannard V, Bensoussan D, Bordigoni P, Baleux F, Le Deist F, Virelizier JL, Arenzana-Seisdedos F, Bachelerie F (2005) WHIM syndromes with different genetic anomalies are accounted for by impaired CXCR4 desensitization to CXCL12. Blood 105: 2449–2457
15. Kramp BK, Sarabi A, Koenen RR, Weber C (2011) Heterophilic chemokine receptor interactions in chemokine signaling and biology. Exp Cell Res 317:655–663
16. Salanga CL, O'Hayre M, Handel T (2009) Modulation of chemokine receptor activity through dimerization and crosstalk. Cell Mol Life Sci 66:1370–1386
17. El-Asmar L, Springael JY, Ballet S, Andrieu EU, Vassart G, Parmentier M (2005) Evidence for negative binding cooperativity within CCR5-CCR2b heterodimers. Mol Pharmacol 67:460–469
18. Sohy D, Parmentier M, Springael JY (2007) Allosteric transinhibition by specific antagonists in CCR2/CXCR4 heterodimers. J Biol Chem 282:30062–30069
19. Sohy D, Yano H, de Nadai P, Urizar E, Guillabert A, Javitch JA, Parmentier M, Springael JY (2009) Hetero-oligomerization of CCR2, CCR5, and CXCR4 and the protean effects of "selective" antagonists. J Biol Chem 284:31270–31279
20. Kenakin T (2011) Functional selectivity and biased receptor signaling. J Pharmacol Exp Ther 336:296–302

21. Contento RL, Molon B, Boularan C, Pozzan T, Manes S, Marullo S, Viola A (2008) CXCR4-CCR5: a couple modulating T cell functions. Proc Natl Acad Sci USA 105:10101–10106
22. Bulenger S, Marullo S, Bouvier M (2005) Emerging role of homo- and heterodimerization in G-protein-coupled receptor biosynthesis and maturation. Trends Pharmacol Sci 26:131–137
23. Hernanz-Falcon P, Rodriguez-Frade JM, Serrano A, Juan D, del Sol A, Soriano SF, Roncal F, Gomez L, Valencia A, Martinez AC, Mellado M (2004) Identification of amino acid residues crucial for chemokine receptor dimerization. Nat Immunol 5:216–223
24. Lemay J, Marullo S, Jockers R, Alizon M, Brelot A (2005) On the dimerization of CCR5. Nat Immunol 6:535, author reply 535–536
25. Hernanz-Falcon P, Rodriguez-Frade JM, Serrano A, Martinez AC, Mellado M (2005) Response to "On the dimerization of CCR5". Nat Immunol 6:535–536
26. Harikumar KG, Pinon DI, Miller LJ (2007) Transmembrane segment IV contributes a functionally important interface for oligomerization of the class II G protein-coupled secretin receptor. J Biol Chem 282: 30363–30372
27. Harikumar KG, Happs RM, Miller LJ (2008) Dimerization in the absence of higher-order oligomerization of the G protein-coupled secretin receptor. Biochim Biophys Acta 1778:2555–2563
28. Gao F, Harikumar KG, Dong M, Lam PCH, Sexton PM, Christopoulos A, Bordner A, Abagyan R, Miller LJ (2009) Functional importance of a structurally distinct homodimeric complex of the family B G protein-coupled secretin receptor. Mol Pharmacol 76:264–274
29. Pioszak AA, Harikumar KG, Parker NR, Miller LJ, Xu HE (2010) Dimeric arrangement of the parathyroid hormone receptor and a structural mechanism for ligand-induced dissociation. J Biol Chem 285:12435–12444
30. Harikumar KG, Ball AM, Sexton PM, Miller LJ (2010) Importance of lipid-exposed residues in transmembrane segment four for family B calcitonin receptor homo-dimerization. Regul Pept 164:113–119
31. Wu B, Chien EY, Mol CD, Fenalti G, Liu W, Katritch V, Abagyan R, Brooun A, Wells P, Bi FC, Hamel DJ, Kuhn P, Handel TM, Cherezov V, Stevens RC (2010) Structures of the CXCR4 chemokine GPCR with small-molecule and cyclic peptide antagonists. Science 330: 1066–1071
32. Manglik A, Kruse AC, Kobilka TS, Thian FS, Mathiesen JM, Sunahara RK, Pardo L, Weis WI, Kobilka BK, Granier S (2012) Crystal structure of the mu-opioid receptor bound to a morphinan antagonist. Nature 485(7398): 321–326
33. Wu H, Wacker D, Mileni M, Katritch V, Han GW, Vardy E, Liu W, Thompson AA, Huang XP, Carroll FI, Mascarella SW, Westkaemper RB, Mosier PD, Roth BL, Cherezov V, Stevens RC (2012) Structure of the human kappa-opioid receptor in complex with JDTic. Nature 485(7398):327–332
34. Wang J, He L, Combs CA, Roderiquez G, Norcross MA (2006) Dimerization of CXCR4 in living malignant cells: control of cell migration by a synthetic peptide that reduces homologous CXCR4 interactions. Mol Cancer Ther 5:2474–2483
35. Lee B, Doranz BJ, Ratajczak MZ, Doms RW (1998) An intricate Web: chemokine receptors, HIV-1 and hematopoiesis. Stem Cells 16:79–88
36. de la Fuente M, Noble DN, Verma S, Nieman MT (2012) Mapping human protease-activated receptor 4 (PAR4) homodimer interface to transmembrane helix 4. J Biol Chem 287:10414–10423
37. Gurevich VV, Gurevich EV (2008) GPCR monomers and oligomers: it takes all kinds. Trends Neurosci 31:74–81
38. Khelashvili G, Dorff K, Shan J, Camacho-Artacho M, Skrabanek L, Vroling B, Bouvier M, Devi LA, George SR, Javitch JA, Lohse MJ, Milligan G, Neubig RR, Palczewski K, Parmentier M, Pin JP, Vriend G, Campagne F, Filizola M (2010) GPCR-OKB: the G protein coupled receptor oligomer knowledge base. Bioinformatics 26:1804–1805
39. Skrabanek L, Murcia M, Bouvier M, Devi L, George SR, Lohse MJ, Milligan G, Neubig R, Palczewski K, Parmentier M, Pin JP, Vriend G, Javitch JA, Campagne F, Filizola M (2007) Requirements and ontology for a G protein-coupled receptor oligomerization knowledge base. BMC Bioinformatics 8:177
40. Hern JA, Baig AH, Mashanov GI, Birdsall B, Corrie JET, Lazareno S, Molloy JE, Birdsall NJM (2010) Formation and dissociation of M1 muscarinic receptor dimers seen by total internal reflection fluorescence imaging of single molecules. Proc Natl Acad Sci USA 107:2693–2698
41. Marullo S, Bouvier M (2007) Resonance energy transfer approaches in molecular pharmacology and beyond. Trends Pharmacol Sci 28:362–365

42. Milligan G, Bouvier M (2005) Methods to monitor the quaternary structure of G protein-coupled receptors. FEBS J 272:2914–2925
43. Ayoub MA, Pfleger KDG (2010) Recent advances in bioluminescence resonance energy transfer technologies to study GPCR heteromerization. Curr Opin Pharmacol 10:44–52
44. Issad T, Jockers R, Ali H, Haribabu B, Walker JM (2006) Bioluminescence resonance energy transfer to monitor protein–protein interactions. In: Walker JM (ed) Transmembrane signaling protocols, vol 332. Humana Press, New York, pp 195–209
45. Achour L, Kamal M, Jockers R, Marullo S, Luttrell LM, Ferguson SSG (2011) Using quantitative BRET to assess G-protein-coupled receptor homo- and heterodimerization. In: Walker JM (ed) Signal transduction protocols, vol 756. Humana Press, New York, pp 183–200
46. Kubale V, Drinovec L, Vrecl M (2012) Quantitative assessment of seven transmembrane receptors (7TMRs) oligomerization by bioluminescence resonance energy transfer (BRET) technology. In: Lapota DD (ed) Bioluminescence—recent advances in oceanic measurements and laboratory applications. InTech, Rijeka, pp 81–95
47. Pfleger KDG, Eidne KA (2006) Illuminating insights into protein–protein interactions using bioluminescence resonance energy transfer (BRET). Nat Meth 3:165–174
48. Kocan M, Pfleger KDG, Willars GB, Challiss RAJ (2011) Study of GPCR–protein interactions by BRET. Receptor signal transduction protocols, vol 746. Humana Press, pp. 357–371
49. Hamdan FF, Percherancier Y, Breton B, Bouvier M (2006) Monitoring protein–protein interactions in living cells by bioluminescence resonance energy transfer (BRET). In Current protocols in neuroscience. Wiley, New York
50. Boute N, Jockers R, Issad T (2002) The use of resonance energy transfer in high-throughput screening: BRET versus FRET. Trends Pharmacol Sci 23:351–354
51. James JR, Oliveira MI, Carmo AM, Iaboni A, Davis SJ (2006) A rigorous experimental framework for detecting protein oligomerization using bioluminescence resonance energy transfer. Nat Meth 3:1001–1006
52. Comps-Agrar L, Maurel D, Rondard P, Pin J-P, Trinquet E, Prézeau L, Luttrell LM, Ferguson SSG (2011) Cell-surface protein–protein interaction analysis with time-resolved FRET and Snap-Tag Technologies: application to G-protein-coupled receptor oligomerization. Signal transduction protocols, vol 756. Humana Press, New York, pp. 201–214
53. Albizu L, Cottet M, Kralikova M, Stoev S, Seyer R, Brabet I, Roux T, Bazin H, Bourrier E, Lamarque L, Breton C, Rives M-L, Newman A, Javitch J, Trinquet E, Manning M, Pin J-P, Mouillac B, Durroux T (2010) Time-resolved FRET between GPCR ligands reveals oligomers in native tissues. Nat Chem Biol 6: 587–594
54. Ferre S, Franco R (2010) Oligomerization of G-protein-coupled receptors: a reality. Curr Opin Pharmacol 10:1–5
55. McLean AJ, Milligan G (2000) Ligand regulation of green fluorescent protein-tagged forms of the human beta(1)- and beta(2)-adrenoceptors; comparisons with the unmodified receptors. Br J Pharmacol 130:1825–1832
56. Mercier J-FO, Salahpour A, Angers SP, Breit A, Bouvier M (2002) Quantitative assessment of beta1- and beta2-adrenergic receptor homo- and heterodimerization by bioluminescence resonance energy transfer. J Biol Chem 277:44925–44931
57. Salahpour A, Masri B (2007) Experimental challenge to a "rigorous" BRET analysis of GPCR oligomerization. Nat Meth 4: 599–600
58. Veatch W, Stryer L (1977) The dimeric nature of the gramicidin A transmembrane channel: conductance and fluorescence energy transfer studies of hybrid channels. J Mol Biol 113:89–102
59. Griesbeck O, Baird GS, Campbell RE, Zacharias DA, Tsien RY (2001) Reducing the environmental sensitivity of yellow fluorescent protein. J Biol Chem 276:29188–29194
60. Patterson G, Day RN, Piston D (2001) Fluorescent protein spectra. J Cell Sci 114:837–838
61. Robitaille M, Héroux I, Baragli A, Hébert TE, Rich PB, Douillet C (2009) Novel tools for use in bioluminescence resonance energy transfer (BRET) assays. Bioluminescence. Vol 574. Humana Press, pp. 215–234
62. Kaczor AA, Selent J (2011) Oligomerization of G-protein-coupled receptors: biochemical and biophysical methods. Curr Med Chem 18:4606–4634
63. Ciruela F, Vilardaga J-P, Fernandez-Duenas V (2010) Lighting up multiprotein complexes: lessons from GPCR oligomerization. Trends Biotechnol 28:407–415

Chapter 8

Unraveling Chemokine and Chemokine Receptor Expression Patterns Using Genetically Engineered Mice

Simon Yona, Ki-Wook Kim, Rebecca Haffner, and Steffen Jung

Abstract

Over the past 25 years, genetically engineered mouse models have become an integral and invaluable research tool to develop our understanding of mammalian physiology and pathology. This unit describes methods for generating transgenic mice, focusing on reporter animals relevant to chemokine receptor and ligand expression. Specifically, we describe the use of bacterial artificial chromosome (BAC) engineering and embryonic stem cell manipulation to generate "*knock in*" and transgenic mice.

Key words Chemokines, Bacterial artificial chromosome, Embryonic stem cells and mice

1 Introduction

1.1 Mononuclear Phagocytes

Over a century ago, Ilya Metchnikoff performed a series of seminal studies spanning from the amoebocyte to the vertebrate macrophage (MØ), to fully comprehend the capabilities of phagocytes [1–4]. Later in the 1960s, Van Furth and Cohn coined the term, the mononuclear phagocyte system [5]. Today the mononuclear phagocyte system is divided into three cell types: circulating monocytes and the more terminally differentiated MØ and dendritic cell (DC), which themselves comprise multiple subpopulations [6].

1.2 Monocyte Subsets and Chemokine Receptors

Circulating monocytes were initially classified by their morphology and density [7–10]. However, these morphological features exhibit a degree of heterogeneity, e.g., variability, nuclear morphology, size, and granularity, making it difficult to discriminate between them by histological criteria alone [11, 12]. Mouse bone marrow, spleen, and circulating monocytes are divided primarily by their chemokine receptor expression and the presence of specific surface molecules. Ly6C and CD115 expression identifies a monocyte subset that expresses high levels of the C–C-chemokine receptor 2 (CCR2) yet intermediate expression of the C-X3-C-chemokine

Astrid E. Cardona and Eroboghene E. Ubogu (eds.), *Chemokines: Methods and Protocols*, Methods in Molecular Biology, vol. 1013, DOI 10.1007/978-1-62703-426-5_8, © Springer Science+Business Media New York 2013

receptor 1 (CX3CR1). In contrast monocytes with low levels of Ly6C express high levels of CX3CR1 and are CCR2 negative [13, 14]. The C–C chemokine receptor 5 is selectively up-regulated on $Ly6C^{lo}$ monocytes that home to atherosclerotic plaques [15].

The C–C chemokine ligands CCL2 and CCL7 bind to CCR2 to recruit $Ly6C^+$ $CCR2^+$ monocytes. Following infection, these C–C chemokine ligands are up-regulated in serum and within inflamed tissue resulting in a substantial quantity of $Ly6C^+$ $CCR2^+$ monocytes egressing from the bone marrow to the peripheral circulation in a CCR2-dependent fashion [13, 16–18]. Therefore, mice immunologically challenged often have increased number of circulating $Ly6C^{hi}$ $CCR2^+$ monocytes.

Chemokine receptor and ligand transgenic mice have helped to elucidate the role of chemokine interactions during cell activation and migration. In addition, mice harboring respective targeted deletions have been instrumental in identifying the importance of chemokine receptors in the orchestration of immune functions as exemplified by the C–C receptor 7 null mouse, which demonstrated CCR7 as a major lymph node-homing receptor on lymphocytes and DCs [19, 20]. Returning to the circulating monocyte, a wealth of data using the CCR2 null mice has shown that these mice display defective $Ly6C^+$ monocyte exit from the bone marrow and a reduced $Ly6C^+$ monocyte recruitment to sites of infection [18].

1.3 $CX_3CR/L1$ and Monocyte Biology

Mice harboring a green fluorescent protein (GFP) reporter in the CX3CR1 locus [21] have been instrumental in identifying and defining multiple mononuclear phagocyte precursors including the macrophage dendritic cell precursor as clonotypic lineage precursors [22] and dedicated DC precursors [23]. Furthermore, these mice have also helped to define tissue-resident mononuclear phagocyte populations [24–26]. Discrete expression levels of CX3CR1/GFP in $CX3CR1^{gfp}$ animals led to the identification of two monocyte subsets in mice as $CX3CR1^{int}$ $Ly6C^{hi}$ and $CX3CR1^{hi}$ $Ly6C^{lo}$ cells [13, 14]. $CX3CR1^{hi}$ $Ly6C^{lo}$ "patrolling" monocytes—the correlate of human $CD14^{dim}$ $CD16^+$ cells [27]—have been shown to adhere and crawl along the luminal vessel surfaces of endothelial cells [28]. $CX3CR1^{int}$ $Ly6C^{hi}$ monocytes are the correlate of human $CD14^+$ $CD16^+$ and $CD14^+$ $CD16^-$ monocytes [27] that are poised to traffic to sites of infection and inflammation [13, 14].

Exploiting the prominent mononuclear phagocyte expression of CX3CR1, we recently generated CX_3CR1^{Cre} mice harboring a *Cre*-recombinase gene in the CX3CR1 locus. In these animals, *Cre*-recombinase is constitutively active in all CX3CR1-expressing cells. Moreover, we also generated $CX3CR1^{CreER}$ mice that carry in their CX3CR1 locus *a Cre*ERT2 gene encoding a fusion protein of the DNA recombinase *Cre* and a mutated ligand-binding domain of the estrogen receptor [29]. This allows the controlled targeting of $CX3CR1^+$ cells and their subsequent progeny. Following tamoxifen administration, normally latent cytoplasmic CreER recombinase becomes liberated and translocates to the nucleus to execute recombination. Copying the $CX3CR1^{gfp}$ locus previously shown to tightly reflect the endogenous CX3CR1 expression pattern, a section of the second CX3CR1 exon is replaced by the Cre or CreERT2 cassette retaining an upstream noncoding exon and extended intron. Comparative analysis of $CX3CR1^{gfp}$, $CX3CR1^{Cre}$:R26-YFP (yellow fluorescent protein), and tamoxifen-treated $CX3CR1^{CreER}$:R26-YFP mice revealed that resident macrophage populations including Kupffer cells and peritoneal macrophages are generated from $CX3CR1^{+}$ precursors prior to birth and maintain themselves during adulthood independent of replenishment by blood monocytes [39].

CX3CL1, the sole ligand of CX3CR1, is synthesized as a type I transmembrane molecule displaying the CX3C chemokine domain on an extended mucin-like stalk [30, 31]. Proteolytic cleavage by the A Disintegrin And Metalloprotease protein, ADAM10, results in constitutive release of a soluble form of CX3CL1 [32]. Moreover, under inflammatory conditions, CX3CL1 shedding is also promoted by ADAM17/TACE (tumor necrosis factor-converting enzyme) [32, 33]. To investigate the in vivo CX3CL1 expression pattern and perform a structure–function analysis, we generated CX3CL1 bacterial artificial chromosome (BAC) transgenic mice [34].

CX3CL1 reporter mice harbor a transgene encoding mCherry reporter under the CX3CL1 promoter. $CX3CL1^{105\Delta}$ mice harbor a BAC transgene encoding only the chemokine domain but lacking the extended mucin-like stalk. Absence of transmembrane domain on these mice allows constitutive secretion of soluble CX3CL1 from the cell surface. These mice elucidated the in vivo expression and physiological function of CX3C chemokine family. CX3CL1 reporter mice demonstrated cell-specific expression in cortical and hippocampal neurons in the brain and goblet cells in the gut epithelium. Analysis of $CX3CL1^{105\Delta}$ mice established the differential roles of shed vs. membrane-bound forms in the formation of intestinal transepithelial dendrites by CX3CR1-expressing gut MØs and blood monocyte survival in vivo [34].

This chapter outlines the basic methods involved in generating transgenic mice. In Protocol 1 we describe the manipulation of BAC to modify genomic DNA regions. Protocol 2 describes the handling and manipulation of embryonic stem (ES) cells.

2 Materials

2.1 Bacterial Artificial Chromosome Manipulation

1. Pulse field gel electrophoresis (PFGE) system: Initial and final switching time = 5 s, run time = 16–20 h, V/cm = 6, included angle = 120°, temperature = 14 °C.

2. Agarose gel: 0.5× Tris–acetate–EDTA (TAE) buffer supplemented with 1 % Megabase Agarose.
3. BAC DNA clones: BAC clones were purchased from the Rosewell Park Cancer Institute library containing our gene of interest. Integrity was verified by PCR analysis and restriction digest followed by PFGE.
4. Luria Broth (LB) agar plates: LB broth and LB-agar plates are supplemented with the appropriate antibiotics to select antibiotic-resistant gene. Tetracyclin (Tet) 3 μg/ml, chloramphenicol (CAP) 15 μg/ml, ampicillin (Amp) 100 μg/ml, and kanamycin (Kan) 50 μg/ml are used as antibiotics.
5. Pipettes: P20, P200, P1000, pipette controller (e.g., Pipetboy).
6. Sterile disposable plasticware: Centrifuge tubes and cryovials. Sterile disposable or reusable detergent-free glass pipettes.
7. Red/ET plasmid kit is purchased from Genebridges.
8. SW106 *E. coli* were purchased from Frederick Nation Lab.
9. BAC DNA isolation kits (Nuclobond BAC100) were purchased from Nucleobond.
10. Biorad MicroPulser™ electroporation apparatus, 0.1 cm electrode gap electroporation cuvette.
11. Erlenmeyer shaker: 500 ml, 1,000 ml.
12. 10 % L-Arabinose stock solution: L-Arabionose (Sigma), dissolved in double-distilled water (DDW) and stored at −20 °C.
13. Thermomixer.
14. Absolute ethanol.
15. PCR tubes.
16. DNA purification kits.
17. Sodium acetate.

2.2 Embryonic Stem Cell Culture and Manipulation

1. Tissue culture: Laminar flow cabinets, humidified incubator (37 °C, 5 % CO_2), inverted phase-contrast microscope, stereomicroscope, centrifuge, vacuum pump, water bath, safety Bunsen burner, tally counter, −70 °C freezer, liquid nitrogen tanks.
2. Sterile plasticware: 35, 60, and 100 mm diameter plates; 4, 6, 24, and 48-well, flat, and U-bottom 96-well plates, centrifuge tubes, and cryovials. Sterile disposable glass pipettes. Sterile disposable reagent reservoirs for a multichannel pipettor and sterile pipette tips.
3. Pipettes: P20, P200, P1000, pipette controller (e.g., Pipietboy) and an electronic multichannel pipettor with volume adjustable up to 200 μl.

4. PFGE with Capacitance Extender Plus, 0.4 cm electrode gap electroporation cuvettes.
5. Proteinase K, 2 mg/ml in water.
6. Geneticin/G418.
7. EmbryoMAX® ES qualified electroporation buffer from Millipore.
8. Trypsin 0.05 %.
9. Phosphate-buffered saline (PBS).

2.3 Tissue Culture Media

Media were prepared and stored at 4 °C for 3–4 weeks.

1. Mouse embryonic fibroblast (MEF) medium: Dulbecco's modified Eagle's medium (DMEM) supplemented with 10 % fetal calf serum (FCS), 50 U/ml penicillin, 50 μg/ml streptomycin, 2 mM L-glutamine, and 1 mM sodium pyruvate.
2. ES cell medium: DMEM supplemented with 15 % FCS (ES cell qualified), 2 mM L-glutamine, 50 U/ml penicillin, 50 μg/ml streptomycin, 0.1 mM nonessential amino acids, 1 mM sodium pyruvate, 0.2 mM beta-mercaptoethanol, and 1,000 U/ml leukemia inhibitory factor (LIF, e.g., ESGRO®).
3. Freezing medium: Freezing medium is prepared fresh prior to use and kept on ice. Mix 80 % ES medium, 10 % FBS, and 10 % dimethyl sulfoxide (DMSO) by volume.

Filter through a 0.2 μm membrane prior to use.

3 Methods

3.1 Bacterial Artificial Chromosome Manipulation

BACs are bacterial vectors that facilitate the stable propagation of large fragments of genomic DNA (100–300 kb), and therefore have a high probability of harboring most regulatory elements of the gene of interest, boundary elements of chromatin domains, or locus control regions, which insulate genes from regulatory influence by neighboring gene [35, 36]. Accordingly, BAC transgenic mice can be considered pseudo-"knock in" mice that allow for a high-throughput analysis of in vivo gene expression. Taking advantage of the modular system and functional cassettes available for BAC cloning, this system may be utilized to generate DNA fragments for standard homologous recombination "*knock in/out*" mice.

BAC transgenesis is a powerful tool, as it allows inclusion of regulatory elements, physiological promoters, and potential undefined boundary elements. The BAC modification relies on a RedE/T recombinant engineering approach (*see* www.genebridges.com), which is based on homologous recombination mediated by the *rec*E and *rec*T proteins referred to as ET-cloning and pBAD-αβγ (expressed from the arabinose-inducible promoter). This is a process whereby *rec*E and *rec*T have been replaced by their respective functional counterparts of phage lambda. *RecE* proteins are 5′→3′ exonucleases. *Rec*T is a DNA-annealing protein. *Rec*E–*rec*T interaction is required for homologous recombination. Short homologous recombination regions to be integrated into circular BAC DNA are designed at the both ends of linear target DNA carrying the neomycin resistance gene required for clone selection following BAC DNA transformation. Short homologous regions (approximately 50–80 nucleotides) are sufficient to promote homologous recombination. Homologous recombination protocol by RedE/T plasmid was modified from Genebridges manual.

3.1.1 Transformation of RedE/T Plasmid into Circular BAC DNA

1. Inoculate 1 ml LB medium (containing chloramphenicol 15 μg/ml) with the BAC clone in *E. coli* and culture overnight at 37 °C.
2. 30 μl of the *E. coli* (BAC clone) culture is added to 1.4 ml of fresh LB broth and culture at 37 °C for a further 1–2 h until optical density of the culture reaches $O.D_{600} = 0.3$.
3. Transfer the *E. coli* (BAC clone) to a centrifuge tube, centrifuge at 12,000 × *g* at 2 °C for 30 s, and discard supernatant.
4. Resuspend the pellet in 1 ml of ice-cold DDW, and pipette up and down ~5 times.
5. Repeat **steps 3** and **4** twice, to completely remove LB broth medium.
6. Following the final centrifugation, add 50 μl of ice-cold DDW to the *E. coli* pellet (final 50 μl) and 100 ng of RedE/T plasmid into *E. coli*. Transfer 50 μl of the *E. coli* (BAC clone)–plasmid mixture into the electroporation cuvette (1 mm).
7. Perform the electroporation with the following suggested settings: 1,350 V, 10 μF, and 600 Ω.
8. Resuspend the *E. coli* (BAC clone) in 1 ml fresh LB medium added to the electroporation cuvette and transfer the mixture into a sterile 1.5 ml centrifuge tube.
9. Allow *E. coli* (BAC clone) to recover at 30 °C for 70 min shaking at 950 rpm.
10. Centrifuge *E. coli* at 12,000 × *g* for 1 min and resuspend *E. coli* in 1 ml LB medium.
11. Plate 100 μl of the resuspended *E. coli* (BAC clone) onto the LB agar plate supplemented with Tet3 and CAP15 antibiotics (*see* Subheading 2.1, **item 4**). Incubate LB agar plate to generate the colony at 30 °C overnight. It is important to maintain at 30 °C but not at 37 °C as RedE/T is temperature sensitive. RedE/T induction by (L)-arabinose will be induced at a later step at 37 °C.

3.1.2 Transformation of Linear Target DNA into Circular BAC DNA

1. Inoculate one colony from *E. coli* transformed with RedE/T plasmid into 1 ml of LB broth (containing Tet3 and CAP15).
2. Culture *E. coli* at 30 °C overnight.
3. Inoculate 30 μl of *E. coli* culture into 1.4 ml of fresh LB broth (containing CAP15) and culture at 37 °C for 2 h until $O.D_{600} = 0.3$, shaking at 1,100 rpm in an orbital/incubating shaker.
4. Add 50 μl of 10 % (L)-arabinose (stock) to induce RedE/T plasmid into the cultured *E. coli.*
5. Culture *E. coli* at 37 °C for a further 45–60 min, shaking at 1,100 rpm.
6. Transfer the bacteria to a centrifuge tube and centrifuge at 12,000 × *g* at 2 °C for 30 s.
7. Resuspend in 1 ml ice-cold DDW and carefully pipette up and down ~5 times.
8. Repeat **steps 6** and 7 twice.
9. Following the final centrifugation, add 20–30 μl of ice-cold DDW to the transformed *E. coli.*
10. Linearize target gene containing homologous regions by restriction with appropriate enzyme.
11. Wash the linearized target gene with a polymerase chain reaction (PCR) purification kit (e.g., from QIAGEN), according to the manufacturer's instructions and quantify.
12. Add 100–200 ng of linearized target gene into the transformed *E. coli*, obtained in **step 9**.
13. Transfer 50 μl of *E. coli*-linearized target gene mixture into an electroporation cuvette (1 mm).
14. Perform the electroporation with the following settings: 1.35 kV, 10 μF, and 600 Ω.
15. Add the 1 ml of fresh LB into the electroporation cuvette.
16. Transfer *E. coli*-LB medium suspension into new 1.5 ml tube.
17. Recover *E. coli* at 37 °C by centrifugation for 3 h shaking at 950 rpm.
18. Centrifuge *E. coli* at 12,000 × *g* for 1 min and resuspend the pellet in 1 ml LB medium.
19. Plate 100 μl of transformed *E. coli* onto the LB agar plates (containing CAP15/Kan15).
20. Incubate plate overnight at 37 °C.

3.1.3 Addition of the Ampicillin Resistance Gene

TARBAC1 backbone plasmid of BAC clone (RP24-147I16) purchased from the Rosewell Park Cancer Institute library has its own wild-type loxP site, which can be replaced with the ampicillin resistance gene by homologous recombination; *see* Subheadings 3.1.1

and 3.1.2. Following homologous recombination, BAC clones with target gene were selected by ampicillin(50), kanamycin(15), and chlorampenicol(15) antibiotics.

3.1.4 Transformation of BAC DNA into SW106 E. coli

BAC DNA with the target gene carrying both ampicllin resistance gene and the neomycin resistance gene flanked by wild-type loxP sites is purified using the NucleoBond® BAC 100 kit and then transformed into SW106 bacteria, in which (L)-arabinose induction activates *Cre* recombinase expression. This feature enables efficient deletion of the neomycin cassette flanked by two *loxP* sites. However, if you intend to use a portion of the BAC for homologous recombination, you may want to leave the neomycin resistance gene in and remove it at a later period.

1. Inoculate one colony from SW106 *E. coli* into 1 ml of LB broth and culture overnight at 37 °C.
2. Transfer 20 μl of SW106 (**step 1**) into 5 ml LB broth in 50 ml tube.
3. Incubate SW106 for 4–5 h at 32 °C until $O.D_{600}=0.5–0.6$.
4. Centrifuge SW106 at 3,200 × *g* for 10 min.
5. Wash SW0106 with 1 ml ice-cold DDW in 1.5 ml centrifuge tube.
6. Centrifuge SW106 at 12,000 × *g* for 1 min.
7. Repeat **steps 5** and **6** twice.
8. Following the final centrifugation, add 20–30 μl of ice-cold DDW to *E. coli* and then add 1–2 ng BAC DNA to *E. coli.*
9. Transfer 50 μl of the *E. coli* BAC DNA mixture into the electroporation cuvette (1 mm).
10. Perform the electroporation with the following suggested settings: 1.75 kV, 25 μF, and 200 Ω.
11. Resuspend the *E. coli* in 1 ml fresh LB medium added to the electroporation cuvette and transfer the mixture into a sterile 1.5 ml centrifuge tube.
12. Allow *E. coli* (BAC clone) to recover at 32 °C for 60 min shaking at 950 rpm.
13. Plate 100 μl of transformed *E. coli* onto the LB agar plates (containing CAP15/Kan15/Amp50).
14. Incubate plate overnight at 37 °C.

3.1.5 Deletion of the Neomycin Resistance Gene

1. Inoculate one colony from BAC DNA-transformed SW106 *E. coli* (BAC-SW106 clone) into 1 ml of LB broth (CAP15/Kan15/Amp50) and culture it at 32 °C overnight.
2. Add 1 ml of BAC-SW106 clone into 10 ml fresh LB broth (CAP15/Kan15/Amp50) and incubate for 2–3 h until $O.D_{600}=0.4–0.5$.

3. Add 100 μl of 10 % (L)-arabionse into 10 ml BAC-SW106 culture and incubate for a further 60 min at 32 °C with 250 rpm shaker.

3.1.6 BAC DNA Purification and Injection

BAC DNA can then be purified from SW106 cells (e.g., using the NucleoBond® BAC 100 kit). BAC DNA was confirmed by PFGE and deletion of neomycin resistance gene was confirmed by PCR. Linearized BAC DNA is ready to be injected into mouse oocytes to create transgenic mice or alternatively introduced into ES cells.

3.1.7 DNA Preparation

1. The DNA is purified with a commercial DNA purification column (e.g., Qiagen). Next, the vector DNA is linearized by restriction enzyme digestion.
2. Precipitate the digested DNA in one-tenth of its volume of 3 M sodium acetate, pH 5.2 and 2.2× its volume of absolute ethanol, and incubate at −20 °C overnight. Spin at 4,000 × *g* for 1 h at 4 °C to recover the DNA.
3. Wash the DNA twice in 500 μl 70 % ethanol and dry briefly.
4. Resuspend DNA in sterile water to a concentration of ~1 μg/μl at 55 °C for 30–60 min. Measure DNA by either UV spectrometer or Nanodrop.

3.2 Embryonic Stem Cell Culture and Manipulation

3.2.1 Growing ES Cells

1. Seed the feeder plate (100 mm diameter), with irradiated (mitotically inactivated) MEFs in MEF medium 1 day before required (*see* **Note 1**).
2. Add 2×10^6 ES cells to the plate in 12 ml ES medium.
3. Change ES medium every 24 h and split the ES colonies at ~80 % confluence (*see* **Note 2**).

3.2.2 Splitting ES Cells

1. Remove the medium, wash cells twice with 10 ml 1× PBS, add trypsin 0.05 % (2 ml per 100 mm diameter plate), and incubate for 5 min or until the cells begin to detach from the plate at 37 °C.
2. Detach residual adherent cells from the bottom of the plate by repetitive pipetting. Add 3 volumes of ES medium to offset the trypsin, pipette up and down ~10 times, and transfer the cell suspension into a 15 ml centrifuge tube. Wash the cells by centrifugation at 300 × *g* for 4 min at room temperature.
3. Remove the supernatant with a laboratory vacuum pump and tap the tube where the pellet is located several times to resuspend the cells before adding 7 ml of ES medium to the tube. Pipette gently to thoroughly mix, and split the contents at a ratio of ~1:5 into the new plates previously plated with MEFs.

3.2.3 Freezing and Thawing ES Cells

ES cells can be frozen like any other mammalian cell.

1. Replace the ES cell medium 3 h before you plan to freeze the cells.
2. Prepare the appropriate amount of freezing medium (*see* Subheading 2.3, **item 3**).
3. Collect the cells in a 15 ml centrifuge tube containing ES medium and pellet the cells as described above in Subheading 3.2.2.
4. Remove the supernatant, then resuspend the cell pellet in 0.5 ml freezing medium (4×10^6 cell/ml), and dispense into cryovials.
5. Place the cryovials in an isopropanol freezing box and leave in –70 °C freezer for 1–2 days, prior to transfer into a liquid nitrogen tank for long-term storage (to enhance cell viability during the freezing).
6. A frozen vial can be thawed directly onto a 100 mm diameter plate. Quickly thaw one vial of ES cells at 37 °C, transfer into a 15 ml centrifuge tube containing 12 ml of ES cell medium, and pellet the cells as described above in Subheading 3.2.2, before resuspending in 12 ml ES medium.

3.2.4 Electroporation

1. Change ES cell medium 2 h prior to electroporation (*see* **Note 3**).
2. Set up electroporation equipment, to 250 V, 500 μF on high capacitance.
3. Place cuvettes and electroporation buffer on ice.
4. Harvest ES cells as described above in Subheading 3.2.2. It is critical to ensure a single-cell suspension. Pool the cells from two 100 mm diameter plates into one 50 ml centrifuge tube.
5. Centrifuge the cells at $300 \times g$ for 6 min. Resuspend in ES cell medium, enumerate cells (e.g., with a hemacytometer), wash, and resuspend the cell pellet to 10×10^6 cells/ml in electroporation buffer. Keep everything at 4 °C from now on.
6. Mix 800 μl of the ES cell suspension with ~30 μg of DNA (*see* Subheading 3.1.5). Transfer the ES cell–DNA mixture into an electroporation cuvette.
7. Electroporate the cells using the settings in **step 2** above. The time constant should be 10 ± 2 ms. Following completion of the electroporation, incubate cuvettes directly on ice for 20 min.
8. Transfer the contents of a cuvette into a 50 ml centrifuge tube containing 30 ml ES medium. Pipette very gently to mix.
9. Remove the medium from premade feeder plates and divide ES cells between six 60 mm diameter plates.

10. Incubate undistributed and change the medium 48 h later.
11. Begin neomycin selection 48 h following electroporation by supplementing the ES medium with 200 μg/ml neomycin analog G418. Change medium every 24 h.
12. Continue the selection process until colonies become clearly conspicuous (this commonly takes between 6 and 10 days).

3.2.5 Picking ES Cell Clones and Screening by PCR for Homologous Recombination

1. Set up a dissecting microscope into an open hood, and organize 96-well plates pipettes and tips systematically.
2. Place 35 μl of trypsin (0.05 %) into each well of a 96-well U-bottom plate.
3. Place the plate on the dissecting microscope. Use a 20 μl pipette set to 10 μl, and scrape a circle around the ES colony to be picked. Detach the MEFs around the colony. Remove the colony from the dish and aspirate it into the tip with as little medium as possible (*see* **Note 4**).
4. Transfer the colony into a well of the trypsin-containing 96-well plate.
5. Ensure you change tip between colonies and fill as many rows as possible within 20 min.
6. Incubate the plate of picked colonies with trypsin at 37 °C for 10 min.
7. Working column by column with a multichannel pipettor, neutralize each colony-containing well with 65 μl of ES medium containing G418. With multichannel pipettor resuspend to a single cell solution.
8. Aspirate the medium from the equivalent columns of premade flat-bottom 96-well plate containing MEFs and transfer the suspension to the correct column, ensuring that the cells are evenly distributed over the entire surface of the well.
9. Wash each well of U-bottom plate with 100 μl of medium to completely remove residual ES cells. Add to the equivalent column in the flat-bottom 96-well plate (total volume of ~200 μl/well).

3.2.6 Screening ES Cells by PCR

1. Remove 20 μl of cell suspension and place it into a clearly labeled PCR tube detailing clone location (i.e., Plate 7 row A and column 1) for PCR analysis later (*see* **Note 5**) (*see* Subheading 3.2.9).
2. Place the plate into the 37 °C incubator.

3. The following day remove half the medium (100 μl) and add fresh ES medium containing G418.
4. Thereafter, completely change the medium daily until the cells are ready for passage (80 % confluence).

3.2.7 Passage and Freezing of ES Cells in 96-Well Plates

1. Aim to split the 96-well plates with two-thirds of each well immediately frozen and one-third transferred to new plate, pre-seeded with fresh MEFs, for a further round of freezing. The second plate serves as an essential backup, in the event that colonies are lost from the first plate.
2. Prepare flat 96-well plates with MEFs at least 4 h before freezing.
3. Prepare 2× freezing medium (*see* Subheading 2.3) and place on ice.
4. Remove the medium from each well and wash each well with 200 μl of 1× PBS using a multichannel pipettor twice.
5. Add 35 μl of trypsin 0.05 % per well. Incubate at 37 °C for 10 min or until the cells begin to detach from the plate observed under an inverted microscope.
6. Add 65 μl of ES medium (without G418) per well to offset the trypsin. With the multichannel pipettor, pipette up and down 15 times to make a single cell solution.
7. Transfer 65 μl of the cell suspension into each well of a U-bottom 96-well plate. Place plate on ice. Working quickly, add 65 μl 2× freezing medium to each well. Pipette up and down to mix.
8. Carefully seal plate with Parafilm and place directly on dry ice for 60 min. Transfer to a −80 °C freezer afterwards.
9. Add 65 μl ES medium per well to the residual cells in the trypsinized plate and transfer to prepared new plate with MEFs. Add another 100 μl of medium to each well of the trypsinized plate, and transfer to the new plate (total volume of ~200 μl/well). Place the plate in the 37 °C incubator.
10. The ES cells should be suitable for a second round of freezing after 2 or 3 days (*see* **Note 6**).

3.2.8 Identifying Correctly Targeted Clones

1. Spin down the PCR tubes containing 20 μl of cells at 4,000 × *g* at 4 °C for 10 min. Remove the supernatant, leaving ~5 μl of medium per tube.
2. Add 15 μl of DDW and boil at 95 °C for 15 min.
3. Add 5 μl of protease K to make a final concentration of 10 μg per tube. Incubate at 55 °C for 90 min.

4. Denature the protease K by boiling the sample at 95 °C for 15 min. The DNA is now ready for the PCR reaction.
5. Add 2× PCR mix to the tube and run a standard RT-PCR to identify positive clones. Note that the expected hit rate can vary; generally we aim to screen around 1,000 clones. In addition, you can expand your colonies up for Southern blot hybridization; this can be performed without feeders, once the cells adapt to these conditions.

Following the identification of positive ES cell lines, these cells should be cultured and expanded for microinjection or aggregation. Your institute ES cell core facility will have specific protocols on how to prepare ES cells for microinjection at this stage.

3.2.9 Confirmation of Germline Transmission

Three weeks following ES cell aggregation, chimeric mice should be born that can be identified according to coat color. During the aggregation process, male ES cells are mixed with unsexed blastocysts. If the host embryo is female and the male ES cells make germ cells, the chimera will often be a fertile male. If the ES cells and host embryo are genetically marked by coat color genes (ES, C57BL/6 vs. blastocyst, B6-albino f.i. B6(Cg)-$^{Tyrc\text{-}2J}$/J), this facilitates assessment of the ES contribution to the animal. Male chimeras can be crossed with an ICR female B6(Cg)-$^{Tyrc\text{-}2J}$/J mice. Black offspring will indicate that ES cells contributed to the germline and transmission of the targeted allele can be assessed by either PCR or southern blot.

4 Notes

1. We grow our ES cells on mitotically inactivated MEFs prepared from PGK neomycin mice. MEF cells can be prepared from any strain of mouse including transgenic mice that express bacterial neomycin. Preparation of MEF stocks is a standard laboratory protocol; for further details, review [37, 38]. Freeze mitotically inactivated MEFs at ~2.2×10^6 cells per vial in 0.5 ml freezing medium. As a rule, we freeze cells slowly and thaw them rapidly. One vial will be sufficient for one 100 mm diameter dish (growth area 58.1 cm^2/dish), two 60 mm diameter dishes (growth area 21.3 cm^2/dish), plates, or one 96-well plate.
2. To preserve the pluripotent potential of ES cells, stringent cell culture conditions should be adhered to. For long-term culture and maintenance of pluripotent ES cells, ES cells are grown on a layer of mitotically inactivated primary MEFs containing LIF and ES medium to promote growth and prevent differentiation. ES cells are kept at a high density and passaged when they reach 70–80 % confluence. Medium should be changed daily. For regular maintenance ES cells should never

be seeded sparsely or grown past 90 % confluence, as these conditions may induce cell differentiation. ES cells should be split ~1:5 depending on their growth rate, every other day (about 1×10^6 ES cells are seeded onto 60 mm diameter plate, 2×10^6 cells onto 100 mm diameter plate). Cells should be trypsinized to form a single-cell suspension as large clumps have a tendency to differentiate. The passage number of ES cells should be kept low and stock and passaged cells should be stored in liquid nitrogen.

3. DNA is introduced into ES cells by electroporation. Passage ES cells 2 days prior to electroporation. Prepare two 100 mm diameter plates grown to ~80 % confluence. This will provide sufficient ES cells for 3 electroporations. Incubate ~30 μg of linearized DNA with 8×10^6 cells per cuvette. Therefore, prepare >100 μg of DNA. The day before electroporation, plate MEFs on 60 mm diameter plates.
4. An ideal plate of ES cell colonies demonstrates well-separated colonies of similar size, with a defined perimeter, a compact center with indistinguishable individual cells. Colonies with this appearance should be picked. Colonies containing flat cells at the edge of the colony are differentiating and should be ignored. Colonies are maintained in positive selection medium until freezing is performed.
5. The day before picking ES colonies, plate flat-bottom 96-well plates with MEFs. Remember, one frozen vial of 2.2×10^6 MEFs is sufficient for one 96-well plate. Using the multichannel pipettor, seed these cells in 200 μl MEF medium. The basic principle is to design PCR primers that will amplify the unique junction between the inserted fragment by homologous recombination and endogenous DNA. This should be performed in advance. The BAC itself acts as a positive control.
6. Keep in mind these time considerations; from generating MEFs to the germline transmission of ES clones it could take about a year. It is important to note that ES cells require daily attention; therefore prior to embarking on any ES cell manipulation (Subheading 3.2.5) until the freezing down your ES clones (Subheading 3.2.8) ensure availability of the required technical expertise to complete the protocol.

References

1. Metchnikoff E (1887) Ueber den Kampf der Zellen gegen Erypselkokken, ein Beitrag zur Phagocytenlehre. Arch Pathol Anat Virchows's Arch 107:209–249
2. Karnovsky ML (1981) Metchnikoff in Messina: a century of studies on phagocytosis. N Engl J Med 304:1178–1180
3. Tauber AI (2003) Metchnikoff and the phagocytosis theory. Nat Rev Mol Cell Biol 4:897–901
4. Gordon S (2008) Elie Metchnikoff: father of natural immunity. Eur J Immunol 38:3257–3264
5. van Furth R, Cohn ZA (1968) The origin and kinetics of mononuclear phagocytes. J Exp Med 128:415–435

6. Yona S, Jung S (2010) Monocytes: subsets, origins, fates and functions. Curr Opin Hematol 17:53–59
7. Tompkins EH (1955) The monocyte. Ann N Y Acad Sci 59:732–745
8. van Furth R, Cohn ZA, Hirsch JG, Humphrey JH, Spector WG, Langevoort HL (1972) Mononuclear phagocytic system: new classification of macrophages, monocytes and of their cell line. Bull World Health Organ 47:651–658
9. Grage-Griebenow E, Flad HD, Ernst M (2001) Heterogeneity of human peripheral blood monocyte subsets. J Leukoc Biol 69:11–20
10. Bennett WE, Cohn ZA (1966) The isolation and selected properties of blood monocytes. J Exp Med 123:145–160
11. Figdor CG, Bont WS, Touw I, de Roos J, Roosnek EE, de Vries JE (1982) Isolation of functionally different human monocytes by counterflow centrifugation elutriation. Blood 60:46–53
12. Yasaka T, Mantich NM, Boxer LA, Baehner RL (1981) Functions of human monocyte and lymphocyte subsets obtained by countercurrent centrifugal elutriation: differing functional capacities of human monocyte subsets. J Immunol 127:1515–1518
13. Geissmann F, Jung S, Littman DR (2003) Blood monocytes consist of two principal subsets with distinct migratory properties. Immunity 19:71–82
14. Palframan RT, Jung S, Cheng G, Weninger W, Luo Y, Dorf M, Littman DR, Rollins BJ, Zweerink H, Rot A et al (2001) Inflammatory chemokine transport and presentation in HEV: a remote control mechanism for monocyte recruitment to lymph nodes in inflamed tissues. J Exp Med 194:1361–1373
15. Tacke F, Alvarez D, Kaplan TJ, Jakubzick C, Spanbroek R, Llodra J, Garin A, Liu J, Mack M, van Rooijen N et al (2007) Monocyte subsets differentially employ CCR2, CCR5, and CX3CR1 to accumulate within atherosclerotic plaques. J Clin Invest 117:185–194
16. Tsou CL, Peters W, Si Y, Slaymaker S, Aslanian AM, Weisberg SP, Mack M, Charo IF (2007) Critical roles for CCR2 and MCP-3 in monocyte mobilization from bone marrow and recruitment to inflammatory sites. J Clin Invest 117:902–909
17. Serbina NV, Salazar-Mather TP, Biron CA, Kuziel WA, Pamer EG (2003) TNF/iNOS-producing dendritic cells mediate innate immune defense against bacterial infection. Immunity 19:59–70
18. Serbina NV, Pamer EG (2006) Monocyte emigration from bone marrow during bacterial infection requires signals mediated by chemokine receptor CCR2. Nat Immunol 7:311–317
19. Forster R, Wolf I, Kaiser E, Lipp M (1994) Selective expression of the murine homologue of the G-protein-coupled receptor BLR1 in B cell differentiation, B cell neoplasia and defined areas of the cerebellum. Cell Mol Biol (Noisy-le-grand) 40:381–387
20. Hopken UE, Droese J, Li JP, Joergensen J, Breitfeld D, Zerwes HG, Lipp M (2004) The chemokine receptor CCR7 controls lymph node-dependent cytotoxic T cell priming in alloimmune responses. Eur J Immunol 34:461–470
21. Jung S, Aliberti J, Graemmel P, Sunshine MJ, Kreutzberg GW, Sher A, Littman DR (2000) Analysis of fractalkine receptor CX(3)CR1 function by targeted deletion and green fluorescent protein reporter gene insertion. Mol Cell Biol 20:4106–4114
22. Fogg DK, Sibon C, Miled C, Jung S, Aucouturier P, Littman DR, Cumano A, Geissmann F (2006) A clonogenic bone marrow progenitor specific for macrophages and dendritic cells. Science 311:83–87
23. Liu K, Victora GD, Schwickert TA, Guermonprez P, Meredith MM, Yao K, Chu FF, Randolph GJ, Rudensky AY, Nussenzweig M (2009) In vivo analysis of dendritic cell development and homeostasis. Science 324: 392–397
24. Niess JH, Brand S, Gu X, Landsman L, Jung S, McCormick BA, Vyas JM, Boes M, Ploegh HL, Fox JG et al (2005) CX3CR1-mediated dendritic cell access to the intestinal lumen and bacterial clearance. Science 307:254–258
25. Varol C, Vallon-Eberhard A, Elinav E, Aychek T, Shapira Y, Luche H, Fehling HJ, Hardt WD, Shakhar G, Jung S (2009) Intestinal lamina propria dendritic cell subsets have different origin and functions. Immunity 31:502–512
26. Bar-On L, Birnberg T, Lewis KL, Edelson BT, Bruder D, Hildner K, Buer J, Murphy KM, Reizis B, Jung S (2010) CX3CR1+ CD8α+ dendritic cells are a steady-state population related to plasmacytoid dendritic cells. Proc Natl Acad Sci USA 107:14745–14750
27. Cros J, Cagnard N, Woollard K, Patey N, Zhang SY, Senechal B, Puel A, Biswas SK, Moshous D, Picard C et al (2010) Human CD14dim monocytes patrol and sense nucleic acids and viruses via TLR7 and TLR8 receptors. Immunity 33:375–386
28. Auffray C, Fogg D, Garfa M, Elain G, Join-Lambert O, Kayal S, Sarnacki S, Cumano A, Lauvau G, Geissmann F (2007) Monitoring of blood vessels and tissues by a population of

monocytes with patrolling behavior. Science 317:666–670

29. Metzger D, Clifford J, Chiba H, Chambon P (1995) Conditional site-specific recombination in mammalian cells using a ligand-dependent chimeric Cre recombinase. Proc Natl Acad Sci USA 92:6991–6995
30. Bazan JF, Bacon KB, Hardiman G, Wang W, Soo K, Rossi D, Greaves DR, Zlotnik A, Schall TJ (1997) A new class of membrane-bound chemokine with a CX3C motif. Nature 385:640–644
31. Fong AM, Erickson HP, Zachariah JP, Poon S, Schamberg NJ, Imai T, Patel DD (2000) Ultrastructure and function of the fractalkine mucin domain in CX(3)C chemokine domain presentation. J Biol Chem 275:3781–3786
32. Hundhausen C, Misztela D, Berkhout TA, Broadway N, Saftig P, Reiss K, Hartmann D, Fahrenholz F, Postina R, Matthews V et al (2003) The disintegrin-like metalloproteinase ADAM10 is involved in constitutive cleavage of CX3CL1 (fractalkine) and regulates CX3CL1-mediated cell-cell adhesion. Blood 102:1186–1195
33. Tsou CL, Haskell CA, Charo IF (2001) Tumor necrosis factor-alpha-converting enzyme mediates the inducible cleavage of fractalkine. J Biol Chem 276:44622–44626
34. Kim KW, Vallon-Eberhard A, Zigmond E, Farache J, Shezen E, Shakhar G, Ludwig A, Lira SA, Jung S (2011) In vivo structure/function and expression analysis of the CX3C chemokine fractalkine. Blood 118: e156–e167
35. Gong S, Yang XW, Li C, Heintz N (2002) Highly efficient modification of bacterial artificial chromosomes (BACs) using novel shuttle vectors containing the R6Kgamma origin of replication. Genome Res 12: 1992–1998
36. Sparwasser T, Gong S, Li JY, Eberl G (2004) General method for the modification of different BAC types and the rapid generation of BAC transgenic mice. Genesis 38:39–50
37. Wurst W, Kuhn R (2008) Gene knockout protocols second edition. Springer, New York
38. Joyner AL (2000) Gene targeting: a practical approach. Oxford University Press, Oxford, p 293, xviii
39. Yona S, Kim KW, Wolf Y, Mildner A, Varol D, Breker M, Strauss-Ayali D, Viukov S, Guilliams M, Misharin A, Hume DA, Perlman H, Malissen B, Zelzer E, Jung S (2013) Fate mapping reveals origins and dynamics of monocytes and tissue macrophages under homeostasis. Immunity 38:79–91

Chapter 9

Zebrafish as a Model to Study Chemokine Function

Eva Kochhan and Arndt F. Siekmann

Abstract

Zebrafish have emerged as a powerful model organism to study embryo morphogenesis. Due to their optical clarity, they are uniquely suited for time-lapse imaging studies, providing insights into the dynamic processes underlying tissue formation and cell migration. These studies have been tremendously facilitated by the availability of transgenic zebrafish lines, labelling distinct embryonic structures, individual cells, or even subcellular structures, such as the nucleus. Zebrafish studies have revealed that the migration of several different cell types in the embryo is controlled by chemokines, small vertebrate-specific proteins. Here, we report methods to analyze the expression pattern of a given chemokine and its receptor in transgenic zebrafish using fluorescent in situ hybridization in combination with an anti-green fluorescent protein (GFP) antibody staining. We furthermore illustrate how to image migrating cell populations using time-lapse microscopy in double-transgenic embryos. We show how to investigate cell number and direction of migration by using a nuclear-localized GFP. The combination of this transgene with a membrane-targeted red fluorescent protein allows for the simultaneous determination of changes in cell shape, such as the formation of filopodial extensions. We exemplify this by describing how a mutation in the chemokine receptor *cxcr4a* affects endothelial cell migration and blood vessel formation. Finally, we provide a method to perform fluorescent angiography to monitor blood vessel perfusion in chemokine receptor mutants.

Key words Zebrafish, Chemokine signalling, Confocal time-lapse imaging, Angiography, Fluorescent in situ hybridization

1 Introduction

Understanding how cells migrate in the embryo to ultimately form different tissues and organs is fundamental to developmental biology. Zebrafish have become an extremely useful model organism for studying mechanisms that influence cell migration. A significant advantage is the transparency of the zebrafish embryo, which allows for detailed cellular and tissue imaging. This feature, together with the rapid embryonic development, makes time-lapse imaging of developmental processes feasible [1, 2]. However, conventional light microscopy has limited applicability in monitoring the movement of

Astrid E. Cardona and Eroboghene E. Ubogu (eds.), *Chemokines: Methods and Protocols*, Methods in Molecular Biology, vol. 1013, DOI 10.1007/978-1-62703-426-5_9, © Springer Science+Business Media New York 2013

individual cells within embryos due to low tissue contrast and the simultaneous detection of light from different focal planes.

A major advancement has been the development of confocal laser microscopy, utilized to detect fluorescently labelled cells and tissues [3]. Initially limited to fixed specimens, the discovery of a whole "fruit basket" of fluorescent proteins in combination with transgenic technologies now permits researchers to follow cells in living embryos with this technique [4, 5]. The ease of generating transgenic zebrafish has provided the scientific community with several hundreds of transgenic zebrafish lines, fluorescently labelling almost every cell type in the body. Recently, transgenic lines have become available that mark distinct subcellular structures, such as the nucleus or the cell membrane of, e.g., an endothelial cell with fluorescent proteins of different colors. Combining these transgenic lines allows for the correlation of cell number, direction of migration, and speed with changes in cell shape, such as the formation of filopodial extensions [6]. An expanding collection of zebrafish mutants and the ability to introduce transmittable lesions in genes of interest via targeting-induced local lesions in genomes (TILLING) or zinc finger nuclease technologies facilitate detailed analysis of gene function in these transgenic lines [7, 8].

One family of proteins influencing cell migration in diverse contexts is the chemokine family, a vertebrate-specific group of small proteins that were initially identified for their role in controlling leukocyte trafficking [9]. Subsequent studies have shown that chemokines and their receptors similarly influence the migration of distinct subsets of cells in the developing embryo [10]. The most extensively studied chemokine/receptor pair is *cxcl12* (*sdf1*)/*cxcr4*. Both play a role during gut vascularization and motor axon pathfinding in mice [11, 12]. Due to a genome duplication event, zebrafish have two *cxcl12* (*cxcl12a* and *cxcl12b*) and *cxcr4* (*cxcr4a* and *cxcr4b*) homologues [13, 14]. *Cxcr4a* is expressed in endothelial cells, and mutants for *cxcr4a* or its ligand *cxcl12b* have defects in the formation of the first embryonic artery and during brain vascularization vascularization [6, 15]. Studies in zebrafish have furthermore elucidated the role of *cxcr4a* in controlling endoderm formation [16]. Zebrafish mutants in *cxcr4b* show impaired germ cell migration [17, 18] and pathfinding defects in olfactory neurons [19]. Together, these studies suggest that chemokine ligand and receptor function during development are likely conserved among vertebrate species.

In order to detect the expression of chemokines or their receptors in distinct cell types, we describe a method that combines fluorescent in situ hybridization (FISH) with anti-green fluorescent protein (GFP) antibody staining in transgenic zebrafish. Embryos stained by this technique can be subsequently analyzed via confocal microscopy, greatly enhancing tissue resolution. This technique furthermore permits the precise co-localization of gene expression with a cell population known to be labelled by the transgene of choice. It circumvents the known limitations of conventional color precipitate-based in situ hybridization (ISH), such as high background or diffusion artifacts. It is also faster and easier to perform than established protocols for two-color FISH [20].

During ISH, the digoxigenin (DIG)-labelled RNA probe is detected with a peroxidase-coupled antibody. Tyramide signal amplification (TSA) is used, allowing for sensitive mRNA detection by enzymatic signal amplification. The secondary anti-peroxidase antibody has a conjugated dye-labelled tyramide and the activation of several tyramides per peroxidase increases the signal within the cells [21]. We show that FISH with combined antibody staining can be used to detect the expression of *cxcr4a* in endothelial cells. However, this protocol can in principle be combined with any transgenic zebrafish line and chemokine/chemokine receptor pair.

In addition, we provide a protocol for performing time-lapse confocal imaging of double-fluorescent, migrating endothelial cells, in which cell nuclei are labelled by green fluorescence and membranes with red mCherry expression. We illustrate how this imaging protocol can be used to determine changes in cell migration upon loss of chemokine receptor function in *cxcr4a* mutant zebrafish. This protocol is also adaptable to different cell types. Finally, we show how to evaluate blood vessel perfusion in chemokine receptor mutants via angiography using Quantum dots. Together, these protocols will help elucidate which chemokine ligands or receptors are expressed in a given tissue and determine the function of these molecules on cell migration.

2 Materials

2.1 Zebrafish Lines and Equipment

To perform time-lapse analysis of blood vessel sprouting in double-transgenic zebrafish with endothelial nuclei expressing GFP and endothelial membranes expressing the red fluorescent protein mCherry, the following transgenic lines are suggested:

1. Tg(*kdrl*:HRAS-mCherry)^{s916} [22]: The promoter of *kdrl* (mammalian equivalent: vascular endothelial growth factor receptor 2 [*vegfr2*]) drives expression of membrane-bound mCherry in all vascular endothelial cells.
2. Tg(*kdrl*:H2B-EGFP)^{mu122} (Kochhan and Siekmann, unpublished): The *kdrl* promoter drives expression of the fusion protein Histone2B-EGFP in the nuclei of all endothelial cells. This can be used to quantify cells in a specific blood vessel.
3. *For* angiography and FISH the following zebrafish line is useful: Tg(*kdrl*:EGFP)^{s843} [23]: The *kdrl* promoter drives expression of a cytoplasmic enhanced green fluorescent protein (EGFP) in all vascular endothelial cells.

4. To analyze the function of the chemokine receptor *cxcr4a* during angiogenesis the following mutant is recommended: *Cxcr4a*^{um21} [6]. This is a zinc finger nuclease-derived mutant containing a premature stop codon in *cxcr4a*, resulting in a truncated protein.
5. Fish tanks with water exchange, tea strainer, and mating boxes with sieves according to your colony setup and husbandry-approved practices.

2.2 Combined Antibody Staining and Fluorescent In Situ Hybridization

1. Plasmid pCRII-cxcr4a [24]: Digest with restriction enzyme *NotI* and use SP6 polymerase for ISH probe generation.
2. DIG-labelled deoxynucleoside triphosphate (dNTP) mix: 10 mM adenosine triphosphate (ATP), cytidine triphosphate (CTP), guanosine triphosphate GTP, 6.5 mM, uridine triphosphate (UTP), and 3.5 mM DIG-11-UTP in distilled water.
3. Respective RNA polymerase with buffer.
4. RNAse inhibitor.
5. DNAse.
6. 0.5 M EDTA.
7. 4 M LiCl.
8. 100 % Ethanol.
9. 1× Phenylthiourea (PTU): 0.003 % PTU in E3 embryo medium (w/v).
10. Phosphate-buffered saline (PBS, 1×): NaCl 137 mM, KCl 2.7 mM, Na_2HPO_4 10 mM, KH_2PO_4 2 mM, pH 7.4 in distilled water.
11. 4 % Paraformaldehyde (PFA) in 1× PBS: Store aliquoted at −20 °C for longer term use (4–6 weeks). Store aliquots at 4 °C when you plan to use up within 1 week.
12. PBS with Tween (PBST): Add Tween-20 to 1× PBS, to make 0.01 % Tween-20.
13. Proteinase K: Produce a stock concentration of 10 mg/ml, store aliquoted at −20 °C.
14. 20× Saline-sodium citrate (20× SSC): Containing 3 M NaCl, 300 mM $Na_3citrate \cdot 2H_2O$, pH 6.0 and store at RT.
15. Hybridization Plus (HYB+) solution: 50 % formamide, 5× SSC, 50 μg/ml heparin, 0.1 % Tween-20, 50 mg/ml torula RNA, store at −20 °C for 2–3 months.
16. 2× Saline-sodium citrate-Tween (SSCT): 0.01 % Tween-20 in 2× SSC.
17. Maleic acid buffer containing Tween-20 (MABT): Prepare solution containing 150 mM maleic acid, 100 mM NaCl, pH 7.5, 0.01 % Tween-20 in distilled water.

18. FISH blocking solution: 2 % blocking reagent (Roche) in MABT, heat until completely dissolved, allow to cool, and store at –20 °C (use within a year).
19. Tyramide signal amplification (TSA) Plus Cyanine 3 (Cy3) solution (Perkin Elmer): Lyophilized tyramide as supplied in kit is dissolved in 60 μl dimethylsulfoxide (DMSO) and store at 4 °C, protected from light, and it is stable for at least 3 months.
20. Peroxidase-coupled anti-DIG antibody (e.g., from Boehringer).
21. GFP-antibody: Rabbit anti-GFP (e.g., from Torrey Pines Bioloabs Inc).
22. Alexa488-anti-rabbit (Ig) antibody (Invitrogen).
23. Glycerol.
24. Slides, coverslips, silicone grease, Watchmaker Forceps (Dumont, size 55).
25. Leica SP5 inverted microscope or confocal microscope as available in your institution.

2.3 Confocal Time-Lapse Imaging

1. E3 embryo medium: 5.07 mM NaCl, 0.16 mM KCl, 0.44 mM $CaCl_2$, 0.68 mM $MgSO_4$ in distilled water and stored at RT for up to 3 months.
2. 1× PTU: 0.003 % PTU in E3 embryo medium (w/v).
3. 1× Tricaine: 168 mg/l Tricaine in E3 embryo medium.
4. Agarose solution: 1 % low-melting-point agarose, 1× PTU, and 1× Tricaine in E3, heat to dissolve, store aliquoted at 4 °C, and melt prior to use.
5. Watchmaker Forceps (Dumont, size 55).
6. Petri dish with glass bottom, for use with an inverted microscope.
7. 6 cm diameter Petri dish for use with an upright microscope.
8. Eyelash to position embryo.
9. Glass Pasteur pipette with holder to transfer embryo into agarose and into Petri dish for embedding.
10. Leica SP5 inverted microscope or confocal microscope as available in your institution.

2.4 Angiography

1. Quantum dots (e.g., Q tracker655 from Invitrogen).
2. Add 4 % methylcellulose (w/v) in E3 medium (e.g., final volume of 100 ml) and store aliquoted at –20 °C.
3. Glass capillary (World Precision Instruments, diameter = 1 mm, internal filament).

4. Glass needles: Prepared from glass capillary by heating the middle and pulling it in two parts using, e.g., the micropipette puller of Sutter Instruments.
5. Pneumatic Picopump (World Precision Instruments): Uses air pressure for injections.

3 Methods

3.1 Zebrafish Husbandry

1. Zebrafish are maintained and mated as described [25]. Briefly, keep fish in the tank at a temperature of 28.5 °C with a light cycle of 14 h. Adult fish require continuous water exchange. Feed once or twice a day with living brine and dried fish food.
2. To mate zebrafish, combine one male and one female in a mating cage without water exchange, separated by a sieve in the evening.
3. The following morning, place the male and female on top of the sieve and switch on the lights to promote mating. Eggs drop through the sieve to the bottom of the mating cage, preventing parental cannibalism.
4. Collect eggs with a tea strainer, transfer to a 10 cm Petri dish containing E3 medium, and place in an incubator maintained at 28.5 °C (without O_2 or CO_2 requirements for the incubator) for up to 5 days.

3.2 Combined Antibody Staining and Fluorescent In Situ Hybridization

In order to determine whether specific genes of interest are expressed in vascular endothelial cells during zebrafish development, we demonstrate how to use FISH for *cxcr4a* in combination with an anti-GFP antibody staining in Tg(*kdrl*:EGFP)^{s843} embryos. In principle, this protocol can be modified for any given gene and transgenic zebrafish line.

For the mRNA detection, a DIG-labelled antisense RNA probe is required. This probe is generated by transcription from plasmids containing either the full-length or partial sequence of the gene of interest (1 kb is sufficient). In addition, the plasmid must contain an RNA polymerase promoter that is located in an antisense orientation. Sp6, T7, or T3 promoters are commonly used.

Figure 1a–c is an illustrative example, demonstrating *cxcr4a* expression in the cells of "nasal ciliary artery" in the head vessels that are also marked by anti-GFP antibody staining in green. Comparison with conventional ISH confirms clear *cxcr4a* localization in vascular endothelial cells of double-labelled embryos.

1. Make the plasmid DNA linear by digesting with a single-cut restriction enzyme (cuts 5′ to the insert). Digest 5 μg DNA with 2 μl enzyme in distilled water in a volume of 100 μl for 2 h. Inactivate the enzyme by heating at the enzyme-specific inactivation temperature for 20 min.

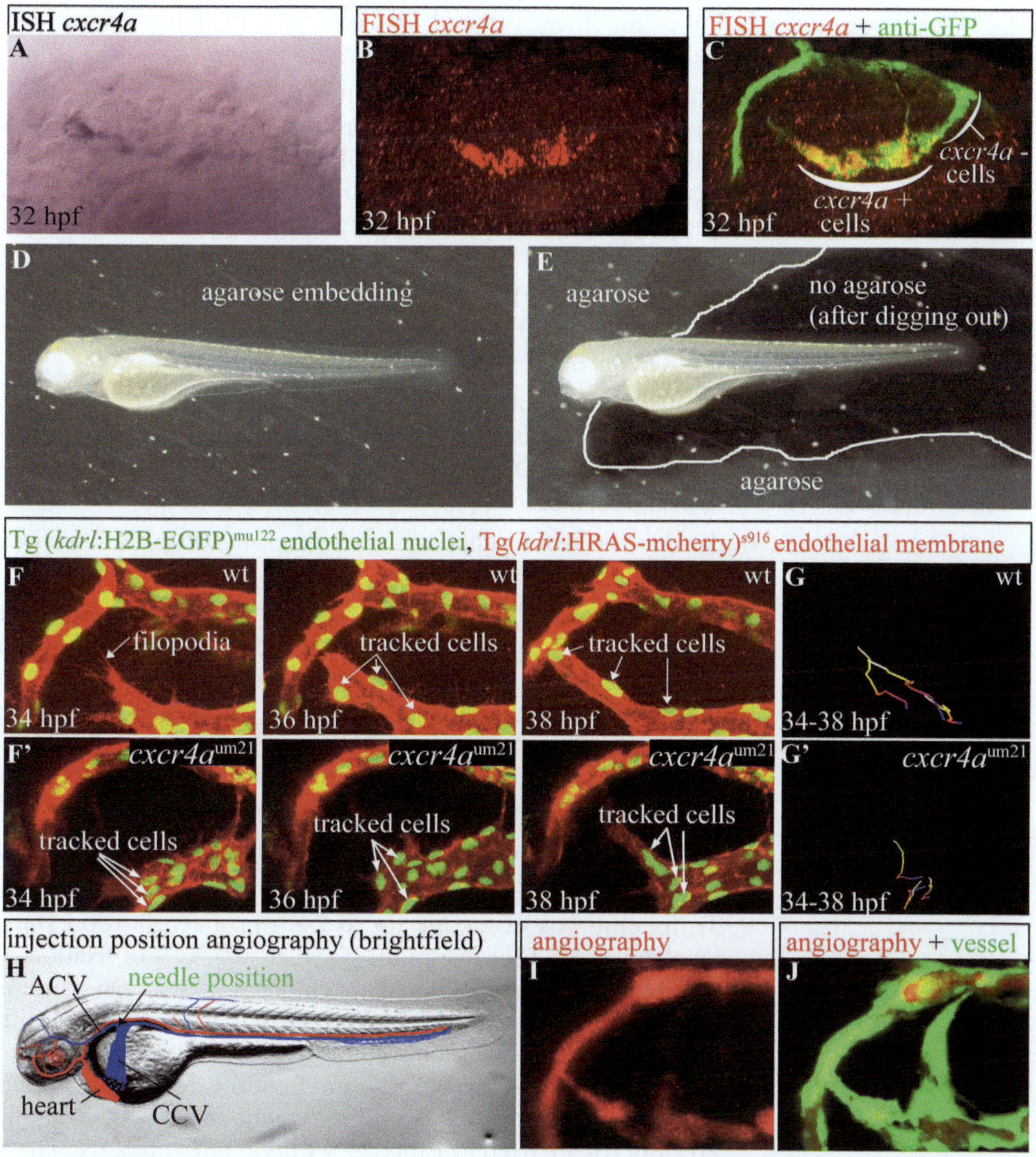

Fig. 1 Application of ISH, confocal imaging, and angiography to analyze chemokine function in zebrafish. (**a**) Comparison of dye precipitate-based ISH and (**b**) FISH for *cxcr4a* expression in head vessels. (**c**) Merge of FISH for *cxcr4a* and anti-GFP antibody-labelled blood vessels in Tg(*kdrl*:EGFP)^{s843} embryos. (**d**) Embryo embedded in agarose for confocal imaging. (**e**) Embryo embedded in agarose after agarose removal to ensure normal development. The agarose surrounding tail, yolk, and heart has been completely removed. (**f**) Confocal time-lapse imaging of Tg(*kdrl*:H2B-EGFP)^{mu122}, Tg(*kdrl*:HRAS-mCherry)^{s916} embryo. Embryos are positioned laterally and head vessels imaged from 34 to 37 hpf to visualize filopodia formation (*labelled with arrow*) and the fusion of two blood vessels. (**f′**) *Cxcr4a*^{um21} mutants are imaged in lateral view from 34 to 48 hpf. (**g**, **g′**) Tracked cell migration of cell nuclei from time-lapse imaging shown in (**f**) or (**f′**), respectively. (**h**) Schematic illustration of selected blood vessels at 48 hpf in lateral view. To perform angiography, the needle is positioned as indicated at the T-shape fusion point where anterior cardinal vein (ACV) and common cardinal vein (CCV) fuse. (**i**) Confocal z-stack of head vessel angiograms at 36 hpf in lateral view. Perfused vessels contain Quantum dots 655. (**j**) Merged projection of EGFP-labelled blood vessels in Tg(*kdrl*:EGFP)^{s843} embryos and associated angiogram

2. Purify the linear plasmid DNA using a PCR purification kit. Elute in 30 μl RNAse-free water.
3. Transcribe the RNA probe by mixing 1 μg linearized DNA, 2 μl DIG-labelled dNTP mix, 2 μl RNA polymerase, 2 μl RNA polymerase 10× buffer, and 1 μl RNAse inhibitor with RNAse-free water in a final volume of 20 μl. Incubate for 2 h at 37 °C. Perform DNAse digestion afterwards by adding 1 μl DNAse for 15 min at 37 °C.
4. Add 0.8 μl 0.5 M EDTA, 2.5 μl 4 M LiCl, and 75 μl 100 % ethanol; mix; and incubate at −20 °C overnight to precipitate the RNA probe.
5. Centrifuge the mixture at 4 °C for 30 min in a microfuge at 13,000 × *g*, aspirate the supernatant, and wash the RNA probe pellet by adding 0.2 ml of 70 % ethanol, gently mixing, centrifuging, and discarding by aspiration the 70 % ethanol.
6. Air-dry the RNA pellet and resuspend in 100 μl RNAse-free water.
7. Remove 5 μl of the probe and run on a 1–1.5 % agarose gel containing 0.5 μg/ml ethidium bromide. Run probe next to 250 ng of uncut plasmid and 250 ng of digested plasmid to determine restriction enzyme efficiency, by the presence of one fragment of expected size for linearized plasmid and the integrity of the RNA probe and by the presence of smaller fragment for RNA probe.
8. Add 400 μl HYB+ to the remaining probe, and store at −20 °C for future use over the next 1–3 years.
9. To fix embryos of an age up to 24 h post fertilization (hpf) incubate in 4 % PFA overnight (ON) at −20 °C. For fixation of older embryonic stages, incubate embryos with a final concentration of 0.003 % PTU (w/v) starting from 24 hpf until fixation. Dechorionate the embryo by gently removing the chorion and expose the embryo and fix at the desired time point (embryonic age) with 4 % PFA overnight at −20 °C. Roughly 20 embryos are fixed in 1 ml of 4 % PFA in one tube ON. These 20 embryos per tube are used for one staining and FISH.
10. Aspirate the fixative, and rinse the embryos twice with PBST. Immediately afterwards, wash embryos by immersion in fresh PBST three times for 5 min each. Dehydrate embryos by incubating in 100 % methanol for 5 min, then transfer to a fresh batch of 100 % methanol, and incubate at −20 °C for at least 30 min (*see* **Note 1**).
11. From here onwards, all steps are carried out at room temperature. Rehydrate embryos by sequential immersion in 75, 50, and 25 % methanol in PBST for 5 min each, followed by washing twice for 5 min in PBST.

Table 1
Proteinase K digestion scheme according to embryo stage for in situ hybridization

Stage	Conc (μg/ml)	Time (min)
4–5 ss	10	1
10–11 ss	10	2
14–15 ss	10	2.5
19–20 ss	10	5
24 hpf	10	13
26 hpf	30	6
28 hpf	30	8
30 hpf	30	10
32 hpf	30	12
36 hpf	30	16
40 hpf	50	12
44 hpf	50	15
48 hpf	50	18
56 hpf	50	22

Embryonic state is indicated as somite stage (ss) or hour post fertilization (hpf). Concentrations of Proteinase K and incubation times used are indicated per stage

12. Digest embryos with Proteinase K as indicated in Table 1, based on somite stage (ss) or hpf.
13. Aspirate Proteinase K, refix embryos with 4 % PFA for 20 min, and then wash three times with PBST for 5 min each. Place embryos in HYB+ solution and incubate at 68 °C for at least 2 h under gentle agitation, in a way that embryos slightly move at the bottom of the tube (pre-hybridization step).
14. Dilute the DIG-labelled RNA probe 1:20 to 1:50 in HYB+ solution. To hybridize roughly 20 embryos per tube, incubate them in 100 μl of diluted probe in a 2 ml reaction tube overnight at 68 °C under gentle agitation (*see* **Note 2**).
15. Aspirate the unbound probe and wash embryos twice in 50 % formamide + 50 % 2× SSCT at 68 °C for 30 min. Wash in 2× SSCT for 15 min, followed by a 30-min wash in 0.2× SSCT.
16. Add 0.5 ml FISH blocking solution to the embryos to block nonspecific binding for at least 1 h at room temperature, followed by antibody incubation (prepared in FISH blocking solution). We recommend a peroxidase-coupled anti-DIG antibody at 1:1,000 dilution and the anti-GFP antibody at

1:50 dilution in blocking solution. Incubate roughly 20 embryos with both antibodies mixed together overnight at 4 °C under agitation in 100 μl volume as needed to ensure complete coverage.

17. Wash embryos four times with 1 ml MABT for 20 min each and twice with 1× PBS for 5 min each.
18. To develop the RNA probe, incubate embryos with 100 μl TSA Plus Cy3 solution (stock tyramide reagent is diluted 1:50 with amplification buffer as provided in the kit) for 60 min in the dark. Wash embryos with PBST three times for 10 min each to remove residual solution.
19. To detect the GFP antibody, block embryos with 500 μl 5 % sheep serum in PBST for 1 h, aspirate the sheep serum, and then incubate for another hour with 100 μl of an Alexa 488-coupled anti-rabbit antibody, diluted 1:500 in PBST containing 5 % sheep serum. Wash three times in PBST for 10 min each.
20. Transfer embryos sequentially into 25, 50, and 75 % glycerol in PBST for 10 min each. Incubate overnight in 75 % glycerol in PBST at 4 °C under agitation to remove unspecific background staining (*see* **Note 3**).
21. Mount embryos in 75 % glycerol in PBST (*see* **Note 4**). In order to prevent the embryo from being squeezed between the microscope slide and the coverslip, apply silicone grease to make four small pillars (approx. 1 mm high). Place an embryo in the middle of these pillars and carefully apply a coverslip on the pillars. Adjust the embryo's orientation by gently pressing the coverslip down.
22. For confocal imaging, we recommend a Leica SP5 inverted microscope with a 20× objective (20× HCX PL S-APO, numerical aperture 0.5, working distance 1.15 mm). Take confocal z-stacks, with a z-step size of 1.5 μm, a pixel size of 1,024 × 512, and a scan speed of 200 Hz. To perform simultaneous two-color imaging, we recommend the "Sequential Scan" function in the Leica confocal software program. Figure 1c shows that *cxcr4a*-positive and -negative endothelial cells within the nasal ciliary artery can be distinguished with this method. This is not possible with conventional ISH.

3.3 Confocal Time-Lapse Imaging to Analyze Chemokine Function During Cell Migration

In order to evaluate the function of a given chemokine or its receptor during cell migration, we carry out confocal time-lapse imaging of wild-type and chemokine receptor mutants using double-transgenic zebrafish. We demonstrate this by showing how to image vascular endothelial cells during angiogenic sprouting and the role of the chemokine receptor *cxcr4a* in this process. To perform this, Tg(*kdrl*:HRAS-mCherry)^{s916}; (*kdrl*:H2B-EGFP)^{mu122} double-transgenic fish are intercrossed in a wild-type or a *cxcr4a* mutant background, respectively.

1. Collect embryos within 5–15 min after eggs are laid (*see* Subheading 3.1 for mating and embryo collection methods) to ensure that embryos at similar stages of development are studied. Two to five hours later, clean embryo clutches by removing dirt and unfertilized eggs using a glass pipette. Place ~50 embryos in a 10 cm diameter Petri dish containing E3 embryo medium, and keep at 28.5 °C. For confocal imaging of embryos older than 24 hpf, incubate in 0.003 % PTU (w/v) in E3 embryo medium to block pigment formation. Otherwise, dark pigments will form and disturb confocal analysis.
2. For time-lapse imaging, a Leica SP5 inverted microscope with a 20× objective (20× HCX PL S-APO, numerical aperture 0.5, working distance 1.15 mm) or a Leica SP5 upright microscope equipped with a 20× objective with dipping lens (20× HCX APO water, numerical aperture 0.5, working distance 3.5 mm) is recommended. Microscopes require a heating stage that should be set to 28.5 °C an hour prior to imaging (*see* **Note 5**).
3. Melt the agarose solution in a heating block maintained at 80 °C. Stop once the agarose solution is completely liquid, and then allow to cool down to 37 °C (*see* **Note 6**).
4. Select embryos at the desired developmental stage. Manually remove the chorion by carefully grabbing the chorion with a pair of fine-tip tweezers and gently pulling the tweezers in opposing directions. Transfer embryos into 1× Tricaine and 1× PTU in E3 embryo medium using a Pasteur pipette in order to immobilize them. Transfer the embryos into the 37 °C agarose solution. Place embryos together with a drop of agarose solution to a Petri dish. To image with an inverted microscope, 40 mm diameter glass bottom dishes are used, while for the upright microscope, 60 mm diameter cell culture dishes are used (*see* **Note** 7).
5. Quickly position the embryo in the dish using an eyelash in an orientation that allows imaging of the blood vessels or the cells or the tissues of interest. Depending on chosen image quality and time frame, you can image 2–3 embryos within one time-lapse by using the "Mark and Find" function of Leica confocal software. To achieve this, mount several embryos close to one another in the same dish without touching. Allow the initial agarose drop (*see* **step 4** above) to dry for 1–2 min. Completely fill the bottom of the dish with 2–4 ml agarose solution, and allow this to dry completely for 5 min.
6. Cover the agarose solution with E3 embryo medium containing 1× PTU and 1× Tricaine to prevent embryos from drying out. Carefully remove the agarose from the top of the heart

and surrounding the yolk and tail using watchmaker forceps. Avoid damaging the embryo. Removing the agarose is important to ensure proper embryo blood flow and growth. Undamaged embryos that exhibit blood flow are processed for confocal imaging analysis (Fig. 1d, e).

7. During time-lapse imaging, take a confocal z-stack every 15 min for 10–16 h. Choose a z-step size between 1.5 and 2.5 μm. We recommend a step size of 1.5 μm, a pixel size of 1,024 × 512, and a scan speed of 200 Hz. Use the "Sequential Scan" function in the Leica confocal software to image two different fluorescent proteins simultaneously. We recommend that you acquire a 100 μm z-stack (approx. 70 steps). With these settings, two embryos can be imaged in parallel during the 15-min time interval between the individual stacks. Embryos need to be covered with E3 embryo medium for the entire duration of time-lapse image acquisition (*see* **Note 8**). Still images from a time-lapse movie showing blood vessel sprouting and fusion as imaged using these settings in double Tg(*kdrl*:HRAS-mCherry)^{s916}; (*kdrl*:H2B-EGFP)^{mu122} are shown in Fig. 1f. Figure 1g contains tracks of migrating endothelial cells, based on nuclear movements. The migratory behavior of *cxcr4a*^{um21} mutant endothelial cells (*see* Fig. 1f′) can be analyzed as well. A comparison of mutant cell migration tracks with wild-type tracks displays less endothelial cell migration in *cxcr4a*^{um21} mutants as shown in Fig. 1g′.

3.4 Angiography

To analyze the influence of chemokine ligand or receptor loss of function on vessel integrity and perfusion, angiography can be performed by injecting small fluorescent molecules into the bloodstream coupled with confocal microscopy. We combine Quantum dots emitting light at 655 nm together with transgenic lines containing GFP-labelled blood vessels, such as Tg(*kdrl*:EGFP)^{s843}. Imaging is carried out using standard laser lines of 561 nm to excite the Quantum dots and 488 nm for GFP excitation (*see* **Note 9**).

1. Mate zebrafish and process embryos as described in Subheadings 3.1 and 3.3. Angiography can be performed after the onset of stable blood flow, i.e., at about 26 hpf. Immobilize a dechorionated embryo in E3 embryo medium containing 1× Tricaine and 1× PTU. Transfer to a drop of methylcellulose placed in a Petri dish. Position the embryo laterally with the left side up in the dish and view under a stereoscope.
2. Load 5 μl of undiluted Quantum dots into a glass needle. Insert the needle into the holder of a picopump. Set the picopump for a timed air pulse of 300 milliseconds (ms). Use a pair of fine forceps to break the tip of the needle, taking care to keep the opening as small as possible.

3. Injections are carried out under brightfield observation and vessels are detected by blood flow. Position the needle at the T-shaped structure where the anterior cardinal vein meets the common cardinal vein, which transports the blood back to the heart as shown in Fig. 1h. The tip of the needle should point towards the head.
4. Push slightly on the back of the holding device to insert the needle into the bloodstream. Keep the injection volume as small as possible between 1 and 10 nl, to minimize effects on cardiac function. This is critical for the success of the angiography procedure (*see* **Note 10**).
5. Immediately after the injection, check the embryo for a normal heartbeat and blood flow in head as well as in the tail under the stereoscope. In addition, you can observe successful Quantum dot injection with a stereoscope equipped with a fluorescent light source. Injected 655 nm Quantum dots are best detected using the DsRed or mCherry filter sets. Mount successfully injected embryos in new methylcellulose or agarose in a Petri dish (*see* Subheading 3.3) and visualize by confocal microscopy (*see* **Note 11**). An example of head region angiography of a 30 hpf-old embryo is shown in Fig. 1i, j.

4 Notes

1. Embryos can be stored in 100 % methanol at −20 °C indefinitely.
2. Most probes generated from plasmids are diluted 1:50. If staining for the DIG-labelled probe is weak, the dilution of the probe can be lowered to 1:20. The RNA probe can be reused up to five times, reducing background staining. Used probe in HYB+ is stored at −20 °C.
3. Exchanging the 75 % glycerol in PBST after 12 h further reduces the background.
4. The Cy3 staining can produce high background signal in the yolk that can sometimes interfere with specific marker detection. In order to eliminate background staining, the yolk should be removed using forceps and an eyelash. Afterwards, embryos are transferred into fresh glycerol prior to imaging.
5. Temperature has a profound effect on the speed of embryonic development. Shifts in temperature can induce developmental abnormalities. Therefore, maintaining a constant temperature of 28.5 °C during the imaging period is of great importance, both for the result reproducibility as well as proper embryonic staging. Different heating chambers are available ranging from simple heat rings that only maintain a constant temperature of the

embryo-containing dish to whole microscope encasings that provide a constant temperature of the embryo in addition to all components in the light path. The latter option is the "gold standard" since it minimizes scanning fluctuations. Nonetheless, we have also successfully used heat rings. When using full microscope encasings, preheating the stage to 28.5 °C an hour prior to imaging ensures that all parts of the microscope and incubation chamber are at the same temperature. The required heating time has to be determined for every system.

6. It is important that the low-melting-point agarose solution is cooled down to ~37 °C for the mounting process to protect the embryo from heat damage.
7. It is important to use cell culture dishes with an appropriate volume of E3 embryo medium when using an upright confocal microscope in order to prevent the embryo from drying out. Evaporation in the dish can be reduced by adding two large surface (plastic) boxes with some water into the heating chamber. Alternatively, E3 embryo medium can be carefully added to the embryo-containing dish between individual scans.
8. Evaporation is minimized at the inverse confocal microscope by closing the dish with a lid.
9. Quantum dots are available that emit light at different wavelengths. Therefore, Quantum dot emission can be adapted to the transgenic zebrafish line expressing the fluorescent protein of choice. In addition, Quantum dot usage can be adapted to the available laser lines. Quantum dots should be spun down at 13,000 × g for 1 min in a benchtop centrifuge prior to use and only 5 μl of the supernatant should be loaded into injection needle.
10. It is better to inject the Quantum dots with up to three small injection doses, instead of a single large injection.
11. After angiography, it is important to exchange methylcellulose and E3 embryo medium in order to eliminate as many Quantum dots that might have leaked out of the needle as possible, as these can cause background signal noise during imaging. The imaging has to be performed within 30 min following injection, since Quantum dots are rather small (15–20 nm in diameter) and can therefore leak out of the blood vessels.

Acknowledgements

We would like to thank Wade William Sugden for critically reading the manuscript. This work was funded by the Max Planck Society, a Deutsche Forschungsgemeinschaft (DFG) grant (SI-1374/3-1), and an ERC starting grant (260794-zebrafishAngio) to AFS.

References

1. Kimmel CB, Ballard WW, Kimmel SR, Ullmann B, Schilling TF (1995) Stages of embryonic development of the zebrafish. Dev Dynam 203:253–310
2. Rieger S, Wang F, Sagasti A (2011) Time-lapse imaging of neural development: zebrafish lead the way into the fourth dimension. Genesis 49:534–545
3. Petroll WM, Jester JV, Cavanagh HD (1994) In-vivo confocal imaging: general-principles and applications. Scanning 16:131–149
4. Shaner NC, Steinbach PA, Tsien RY (2005) A guide to choosing fluorescent proteins. Nat Methods 2:905–909
5. Kawakami K, Koga A, Hori H, Shima A (1998) Excision of the Tol2 transposable element of the medaka fish, Oryzias latipes, in zebrafish, Danio rerio. Gene 225:17–22
6. Bussmann J, Wolfe SA, Siekmann AF (2011) Arterial-venous network formation during brain vascularization involves hemodynamic regulation of chemokine signaling. Development 138:1717–1726
7. Meng XD, Noyes MB, Zhu LHJ, Lawson ND, Wolfe SA (2008) Targeted gene inactivation in zebrafish using engineered zinc-finger nucleases. Nat Biotechnol 26:695–701
8. Winkler S, Gscheidel N, Brand M (2011) Mutant generation in vertebrate model organisms by TILLING. Methods Mol Biol 770:475–504
9. Esche C, Stellato C, Beck LA (2005) Chemokines: key players in innate and adaptive immunity. J Invest Dermatol 125: 615–628
10. Raz E, Mahabaleshwar H (2009) Chemokine signaling in embryonic cell migration: a fisheye view. Development 136:1223–1229
11. Tachibana K, Hirota S, Iizasa H, Yoshida H, Kawabata K, Kataoka Y, Kitamura Y, Matsushima K, Yoshida N, Nishikawa S, Kishimoto T, Nagasawa T (1998) The chemokine receptor CXCR4 is essential for vascularization of the gastrointestinal tract. Nature 393:591–594
12. Lieberam I, Agalliu D, Nagasawa T, Ericson J, Jessell TM (2005) A Cxcl12-Cxcr4 chemokine signaling pathway defines the initial trajectory of mammalian motor axons. Neuron 47: 667–679
13. Kiefer F, Siekmann AF (2011) The role of chemokines and their receptors in angiogenesis. Cell Mol Life Sci 68:2811–2830
14. Chong SW, Emelyanov A, Gong ZY, Korzh V (2001) Expression pattern of two zebrafish genes, cxcr4a and cxcr4b. Mech Develop 109:347–354
15. Siekmann AF, Standley C, Fogarty KE, Wolfe SA, Lawson ND (2009) Chemokine signaling guides regional patterning of the first embryonic artery. Genes Dev 23:2272–2277
16. Nair S, Schilling TF (2008) Chemokine signaling controls endodermal migration during zebrafish gastrulation. Science 322:89–92
17. Doitsidou M, Reichman-Fried M, Stebler J, Koprunner M, Dorries J, Meyer D, Esguerra CV, Leung T, Raz E (2002) Guidance of primordial germ cell migration by the chemokine SDF-1. Cell 111:647–659
18. Knaut H, Werz C, Geisler R, Nusslein-Volhard C, Consortium, T. S. (2003) A zebrafish homologue of the chemokine receptor Cxcr4 is a germ-cell guidance receptor. Nature 421:279–282
19. Miyasaka N, Knaut H, Yoshihara Y (2007) Cxcl12/Cxcr4 chemokine signaling is required for placode assembly and sensory axon pathfinding in the zebrafish olfactory system. Development 134:2459–2468
20. Julich D, Hwee Lim C, Round J, Nicolaije C, Schroeder J, Davies A, Geisler R, Lewis J, Jiang YJ, Holley SA (2005) Beamter/deltaC and the role of Notch ligands in the zebrafish somite segmentation, hindbrain neurogenesis and hypochord differentiation. Dev Biol 286: 391–404
21. Clay H, Ramakrishnan L (2005) Multiplex fluorescent in situ hybridization in zebrafish embryos using tyramide signal amplification. Zebrafish 2:105–111
22. Hogan BM, Bos FL, Bussmann J, Witte M, Chi NC, Duckers HJ, Schulte-Merker S (2009) Ccbe1 is required for embryonic lymphangiogenesis and venous sprouting. Nat Genet 41: 396–398
23. Jinn SW, Beisl D, Mitchell T, Chen JN, Stainier DYR (2005) Cellular and molecular analyses of vascular tube and lumen formation in zebrafish. Development 132:5199–5209
24. Fujita M, Cha YR, Pham VN, Sakurai A, Roman BL, Gutkind JS, Weinstein BM (2011) Assembly and patterning of the vascular network of the vertebrate hindbrain. Development 138:1705–1715
25. Westerfield M (1993) The zebrafish book. University of Oregon Press, Eugene, OR

Chapter 10

Inflammatory Macrophage Migration in Experimental Autoimmune Encephalomyelitis

William J. Karpus

Abstract

Experimental autoimmune encephalomyelitis (EAE) is a CD4 T cell-mediated demyelinating disease of the central nervous system (CNS) where macrophages are the end-stage effector cell. EAE serves as a model for multiple sclerosis where it has been instructive in delineating the autoimmune cellular response in the CNS for the purpose of developing more effective therapies. Understanding the nature of how cytokine and chemokine networks regulate the migration of leukocytes to the CNS requires the ability to track subpopulations of those cells in vivo. We describe a flow cytometric technique to monitor the migration of macrophages during EAE development.

Key words Multiple sclerosis, Macrophage, Central nervous system, Autoimmune encephalomyelitis, Chemokines, Flow cytometry

1 Introduction

Experimental autoimmune encephalomyelitis (EAE) is a CD4 T cell-mediated demyelinating disease of the central nervous system (CNS) that serves as an animal model for the human disease multiple sclerosis (MS) [1]. Autoreactive T cell and macrophage migration to the CNS is a necessary component of disease pathogenesis [2, 3]. In fact, one of the currently approved therapies for MS, natalizumab, functions by blocking the α4β1-mediated T cell entry into the CNS [4, 5]. A large body of evidence suggests that chemokines control the migration of leukocytes to the CNS during the development and progression of EAE as well as MS [6, 7]. In fact, genetic deletion of CCR1 [8], CCR2 [9], CCR4 [10], and CXCR2 [11] have indicated that these molecules and responsiveness to their ligands [12–14] are important in disease development. These observations suggest that small-molecular-weight chemokine receptor antagonists can be developed as therapeutics. Critical to understanding how leukocyte subsets migrate to the

Astrid E. Cardona and Eroboghene E. Ubogu (eds.), *Chemokines: Methods and Protocols*, Methods in Molecular Biology, vol. 1013, DOI 10.1007/978-1-62703-426-5_10, © Springer Science+Business Media New York 2013

CNS is the ability to complement targeted genetic deficiencies by adoptive transfer of specific leukocyte subpopulations that express the deleted gene and the ability to identify those populations in vivo [15].

We recently described a role for CCR4 [10] and its ligand CCL22 [14] in the development of EAE. Our data suggest that genetic deletion of CCR4 resulted in attenuated disease development from less inflammatory macrophage accumulation in the CNS. Similarly, in vivo neutralization of the CCR4 ligand, CCL22, also resulted in less disease severity. However, in order to unequivocally demonstrate that the inflammatory macrophage ($CD45^{high}CD11b^{+}CD11c^{-}Ly6C^{high}$; iMφ) is the leukocyte subpopulation attracted to the CNS by CCL22 and is an effector cell involved in EAE development, adoptive reconstitution of CCR4-deficient mice with a wild-type, CCR4-expressing iMφ population and the ability to detect it in the CNS is a necessary approach. To that end we describe a flow cytometric technique using the allelic difference of CD45 (CD45.1 vs. CD45.2) as a method to visualize adoptively iMφ that migrated into the CNS.

2 Materials

2.1 Mice

1. Wild-type C57BL/6 ($CD45.2^{+}$) mice (e.g., The Jackson Laboratory).
2. C57BL/6 CD45.1 congenic mice [16]: *B6.SJL-Ptprca Pepcb/BoyJ* mice are available from The Jackson Laboratory.
3. C57BL/6 $CCR4^{-/-}$ mice [17]: *B6;129P-Ccr4tm1Pwr/J* mice are available from The Jackson Laboratory.

2.2 Macrophage Isolation Reagents

1. 3 % Thioglycollate solution: Weigh powdered thioglycollate broth medium and carefully add to sterile distilled H_2O in a beaker to make a 3 % weight/volume (w/v) solution. Bring solution to boil with stirring and autoclave to sterilize (*see* **Note 1**).
2. Macrophage harvesting medium: Add sterile, endotoxin-free fetal bovine serum (FBS) to Dulbecco's minimum essential medium (DMEM) to make a 5 % volume/volume (v/v) solution.
3. 15 and 50 ml conical polypropylene centrifuge tubes (*see* **Note 2**).
4. 18-, 25-, and 30-gauge needles.
5. 1, 10, and 30 ml polypropylene syringes.
6. Percoll® (e.g., Sigma-Aldrich).
7. 1× and 10× Hank's Balanced Salt Solution (HBSS).

2.3 Flow Cytometry Reagents/Instrumentation/Software

1. Flow buffer: 1 % (w/v) bovine serum albumin (BSA, fraction V) in 0.2 μm filtered 1× phosphate-buffered saline (PBS) pH 7.2. To make 1× PBS, dissolve 0.23 g anhydrous NaH_2PO_4 (final concentration 1.9 mM), 1.15 g anhydrous Na_2HPO_4 (final concentration 8.1 mM), and 9.0 g NaCl in 900 ml MilliQ distilled, deionized water (ddH_2O). Add 10 g BSA to the 1× PBS with continuous stirring. Adjust pH using 1 M HCl and then add ddH_2O to achieve a final volume of 1 L.
2. Antibodies with clone designation in parentheses: CD45-PerCP (30-F11), CD11b-APC-Cy7 (M1/70), CD11c-APC (HL3), Ly6C-FITC (AL-21), CD45.1-APC (A20), CD45.2-PerCP-Cy5.5 (104), and CCR4-PE (2 G12). All antibodies were purchased from BD Biosciences except CCR4, which was purchased from Biolegend (*see* **Note 3**).
3. Analytical flow cytometer, e.g., BD Canto II instrument equipped with 488 nm solid-state and 633 nm HeNe lasers. The instrument setup and spectral compensation should be performed as previously described [18].
4. Flow cytometry software, e.g., Data acquisition (FACSDiva, BD Biosciences) and analysis (FlowJo, Treestar) (*see* **Note 4**).

3 Methods

3.1 Macrophage Isolation

1. Inject 1 ml of the 3 % thioglycollate solution using a 25-gauge needle and 10 ml syringe into the peritoneal cavity of donor wild-type C57BL/6.
2. Euthanize mice 7 days later using locally approved methods (*see* **Note 5**). Surgically expose the peritoneal wall by making a midline abdominal incision of only the skin with fine dissecting scissors, leaving the peritoneal membrane intact.
3. Inject 10 ml of the macrophage-harvesting medium into the peritoneal cavity using an 18-gauge needle attached to a 30 ml syringe.
4. Aspirate the peritoneal leukocyte-containing fluid by withdrawing slowly with the same needle and syringe as in **step 3** above. A harvest of 5–10 ml of peritoneal fluid can be expected from each mouse (*see* **Note 6**).
5. Pool the macrophage-containing peritoneal fluid in a 50 ml conical tube. Centrifuge at 500 × *g* for 15 min at 4 °C. Completely aspirate the supernatant and resuspend the cell pellet in 15 ml 1× HBSS. Wash by repeating centrifugation step as described earlier and resuspend in 1× HBSS.
6. Quantify isolated cells with a hemacytometer. Adjust concentration to 10^7 cells/ml in 1× HBSS (*see* **Note 7**) and store on

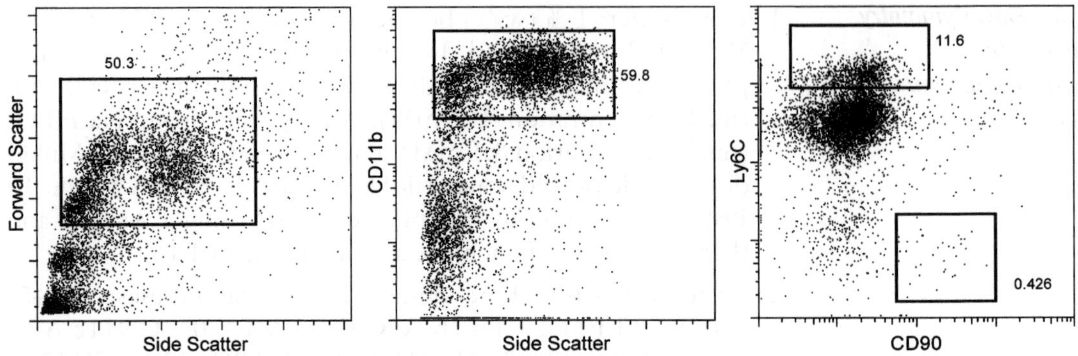

Fig. 1 Flow cytometric assessment of iMφ content in an elicited peritoneal leukocyte population. Seven days after intraperitoneal injection of thioglycollate broth, mice were euthanized and the stimulated leukocytes were harvested and immunophenotypically assessed by flow cytometry. The gating strategy is as follows: Events were selected using forward versus side scatter to eliminate debris (*left panel*); from that specific gate, $CD11b^+$ cells plotted versus side scatter are shown (*middle panel*) and a gate encompassing the bright $CD11b^+$ events was established as this is a molecule expressed by macrophages; the events in the CD11b gate are then displayed as $Ly6C^+$ versus $CD90^+$ events (*right panel*). The numbers refer to the percentage of events in each specific gate and indicate the peritoneal leukocyte population consisted of 11.6 % iMφ and <1 % contaminating T cells

ice during flow cytometric analysis. Bring the cells to room temperature prior to adoptive transfer.

7. Aliquot 10^6 washed peritoneal leukocytes (100 μl) to determine the percentage of iMφ by flow cytometric analysis. Bring the volume of 1× HBSS to 15 ml and centrifuge at 500 × *g* for 15 min at 4 °C. Discard the supernatant and resuspend the pellet in a 0.1 ml antibody cocktail in flow buffer. The cocktail contains dilutions of the antibodies used to identify and quantify iMφ (*see* **Note 8**).
8. Incubate leukocytes in the antibody cocktail for 30 min on ice in the dark. Wash as described in **step 7**, substituting flow buffer for 1× HBSS. Resuspend cell pellet in 1 ml flow buffer.
9. Determine the percentage of iMφ in the overall peritoneal leukocyte population by flow cytometry based on the following immunophenotype: $CD45^{high}CD90^-CD11b^+CD11c^-Ly6C^{high}$ (*see* **Note 9**). A typical result is shown in Fig. 1.

3.2 Adoptive iMφ Transfer

1. Calculate the number of iMφ in the sample using the total washed peritoneal leukocyte count as described above (*see* Subheading 3.1, **step 6**) multiplied by the percentage of $CD45^{high}CD11b^+CD11c^-Ly6C^{high}$ cells as determined by flow cytometry (*see* **Note 10**).
2. Intravenous iMφ transfer: Inject 0.2 ml of the cell suspension in 1× HBSS into the donor mouse lateral tail vein using a 30-gauge needle attached to a 1 ml syringe. Mice are placed in a manual restraining tube (e.g., from VWR) prior to cell transfer. The optimal HBSS volume for cell transfer is 0.1–0.2 ml and

the number of adoptively transferred cells necessary to observe the desired effect must be determined empirically. However, in our experience the range of cell number is 1–10 × 10^6 (*see* **Note 11**).

3.3 Analytical Flow Cytometry

In order to ascertain whether a population of chemokine receptor expressing iMφ would modulate the severity of EAE observed in the mutant strain (e.g., CCR4), EAE can be induced in wild-type and chemokine receptor-deficient mice, as previously described [19]. On day 7 following immunization, adoptive transfer of wild-type iMφ (CD45.1$^+$) to the chemokine receptor-deficient mice (CD45.2$^+$) is performed (*see* Subheading 3.2, **step 2**).

1. To determine whether the adoptively transferred iMφ migrated to the CNS during EAE, euthanize the recipient mice at 14 days post disease induction by CO_2 asphyxiation.
2. Dissect/harvest the spinal cord from each mouse and homogenize in DMEM (*see* **Note 10**). Euthanize mice as previously described and dissect the spinal cord from the vertebral column by careful excision using fine scissors so as not to disrupt the meninges. Pool the spinal cords in 1× HBSS and homogenize through a 100 μm mesh wire screen using a 2 ml disposable syringe plunger. Wash the cell suspension in 1× HBSS as previously described (*see* Subheading 3.1, **step 7**) and the cell pellet is resuspended in 5 ml 1× HBSS.
3. Extract leukocytes from the spinal cord homogenate using a discontinuous Percoll® gradient as previously described [20]. Stock solutions of 30 and 70 % Percoll® are prepared by diluting 100 % Percoll® with 10× HBSS. The Percoll® gradient is prepared by layering 5 ml of 70 % Percoll® at the bottom of a 15 ml conical centrifuge tube, carefully adding 5 ml 30 % Percoll® on top of the 70 % layer (each layer of Percoll® must be layered slowly as it is important to avoid mixing between the interfaces in order to obtain optimal cell separation efficiency), and finally layering 5 ml of the cell-containing 1× HBSS solution on top of the 30 % Percoll layer. Centrifuge the tubes at 500 × *g* for 15 min at 4 °C as previously described without using a rotor brake. Carefully aspirate the cellular layer at the 70–30 % Percoll® layer interface taking as little Percoll® as possible while capturing as many cells as possible. Wash the recovered cells in 1× HBSS, bringing the volume up to 50 ml to sufficiently dilute any aspirated Percoll.

   Stain with an antibody cocktail specific for CD45.1, CD45.2, CD11b, CD11c, chemokine receptor, and Ly6C and perform flow cytometry data acquisition and analysis (*see* Subheading 3.1, **steps 8** and **9**).
4. Distinguish adoptively transferred wild-type iMφ (CD45.1$^+$) that migrated to the CNS during EAE development from host-derived macrophages (CD45.2$^+$) based on CD45 staining characteristics, as shown in Fig. 2 (*see* **Note 11**).

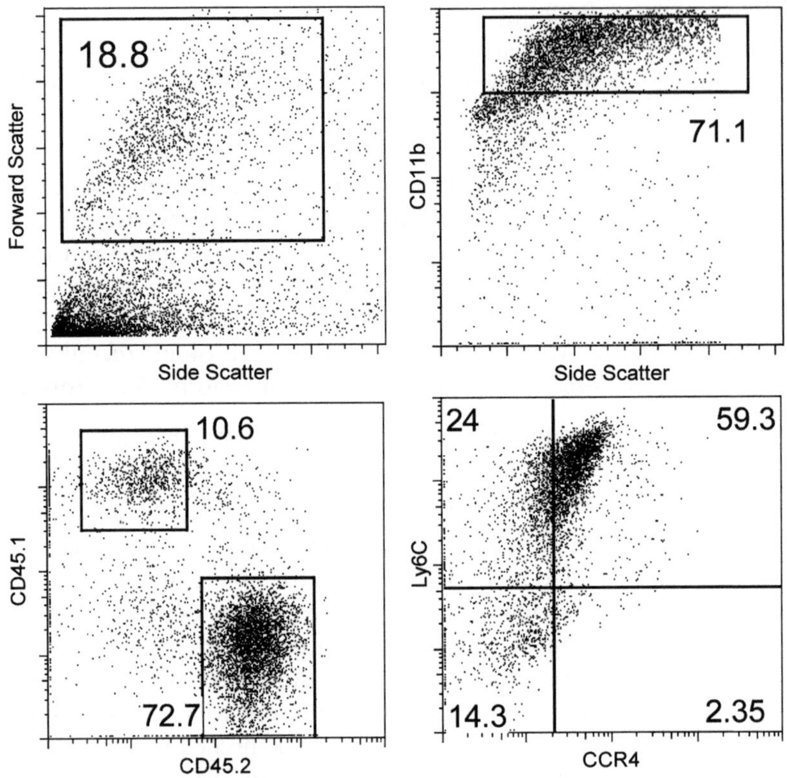

Fig. 2 CCR4$^+$ iMφ migrate to CNS during EAE. Fourteen days after disease induction and 7 days after adoptive iMφ transfer, mice were euthanized and spinal cord-infiltrating leukocytes were isolated and immunophenotypically examined by flow cytometry. The gating strategy used in this particular analysis was as follows: Events were selected using forward versus side scatter to eliminate debris inherent in the CNS leukocyte isolation procedure (*upper left panel*); from that specific gate, CD11b$^+$ cells plotted versus side scatter are shown (*upper right panel*) and a gate encompassing the bright CD11b$^+$ events was established as this is a molecule expressed by macrophages (alternatively the marker F4/80 can also be used); the events in the CD11b gate are then displayed as CD45.1$^+$ versus CD45.2$^+$ events (*lower left panel*); since adoptively transferred cells were derived from a CD45.1$^+$ mouse and the recipients' cells only express CD45.2, the cells identified as CD45.1$^+$CD45.2$^-$ are the iMφ; the CD45.1$^+$ events are then verified as Ly6C^{high}CCR4$^+$ (*lower right panel*). The numbers refer to the percentage of events in each specific gate

4 Notes

1. Store at room temperature. Discard if solution becomes turbid.
2. Polystyrene tubes can also be used; however, cell clumping, sticking, and loss are minimized with polypropylene tubes.
3. Antibodies to cell surface markers can be purchased from any vendor, as we have found no differences among antibodies from various suppliers. We have found that the anti-CCR4 antibody is only available from Biolegend.

4. FACS Diva was used as the acquisition software because it came installed on the flow cytometer workstations.
5. Methods of euthanasia such as CO_2 asphyxiation are preferred over methods such as intraperitoneal injection of ketamine–xylazine as the latter may result in significant blood contamination.
6. Within the harvested peritoneal fluid, there should be approximately 10^7 total macrophages per mouse.
7. Enumerate macrophages using a hemacytometer and a light microscope. Make sure that macrophages are in complete suspension. Draw 0.01 ml (10 μl) of the suspension into a micropipette and dispense between the coverslip and the hemacytometer. Focus the light microscope on the cells and count the number of cells in each of the four corner squares. Count the number of cells as follows: mean number of cells/0.1 mm^3 volume = total cells in four corner squares/4. Total number of cells per ml (cells/cm^3) = (mean number of cells/0.1 mm^3 volume) × 10^4.
8. The concentration of antibody used to stain the peritoneal leukocyte population should be determined empirically to provide the best signal-to-noise ratio [21]. It is our experience that the indicated antibodies used in the present analysis provide robust signal-to-noise ratios at a 1:100 dilution.
9. CD90 is used as a T cell marker to assist in the flow cytometric gating strategy and determine the level of lymphocyte contamination in the preparation. In addition to the immunophenotype, determination of live vs. dead cells can be performed by flow cytometry using a DNA-intercalating dye such as propidium iodide or 7-aminoactinomycin D (7-AAD).
10. As an alternative to transferring the isolated CD45.1$^+$ leukocytes containing a subpopulation of iMφ to CD45.2$^+$ recipient mice, iMφ can be cell sorted by flow cytometry and adoptively transferred afterwards [22]. The cell sorting procedure is similar to analytical flow cytometry with the exception that the cells are deposited into a centrifuge tube, washed, and resuspended to the desired concentration prior to transfer to recipient mice.
11. The number of cells transferred to recipient mice should be empirically determined. As an alternative to intravenous injection, intraperitoneal injection can be used. However, it is our experience that intravenous injection provides a more reliable and consistent route for introducing iMφ into the circulation.
12. 1× HBSS can be substituted for DMEM.
13. The gating strategy used in this particular analysis was as follows: Select events using forward versus side scatter to

eliminate debris inherent in the CNS leukocyte isolation procedure. From that specific gate, display the results as CD11b versus side scatter and establish a gate encompassing the $CD11b^+$ cells, as this is a molecule expressed by macrophages (alternatively the marker F4/80 can also be used). The events in the CD11b gate are then displayed as $CD45.1^+$ versus $CD45.2^+$ events. Adoptively transferred cells were derived from a $CD45.1^+$ mouse and the recipients' cells only express CD45.2; hence the cells identified as $CD45.1^+CD45.2^-$ are the iMφ; the $CD45.1^+$ events are then verified as chemokine receptor +, $Ly6C^{high}$.

Acknowledgements

This work was supported by National Institutes of Health grant NS34510 and NMSS grant RG4146.

References

1. Steinman L, Zamvil SS (2006) How to successfully apply animal studies in experimental allergic encephalomyelitis to research on multiple sclerosis. Ann Neurol 60:12–21
2. Hickey WF, Hsu BL, Kimura H (1991) T-lymphocyte entry into the central nervous system. J Neurosci Res 28:254–260
3. Huitinga I, Van Rooijen N, De Groot CJA et al (1990) Elimination of macrophages infiltrating into the central nervous system suppresses experimental allergic encephalomyelitis in Lewis rats. Ann N Y Acad Sci 594:458–460
4. Yednock TA, Cannon C, Fritz LC et al (1992) Prevention of experimental autoimmune encephalomyelitis by antibodies against α4β1 integrin. Nature 356:63–66
5. Baron JL, Madri JA, Ruddle NH et al (1993) Surface expression of alpha 4 integrin by CD4 T cells is required for their entry into brain parenchyma. J Exp Med 177:57–68
6. Sorensen TL, Tani M, Jensen J et al (1999) Expression of specific chemokines and chemokine receptors in the central nervous system of multiple sclerosis patients. J Clin Invest 103:807–815
7. Mahad DJ, Ransohoff RM (2003) The role of MCP-1 (CCL2) and CCR2 in multiple sclerosis and experimental autoimmune encephalomyelitis (EAE). Semin Immunol 15:23–32
8. Rottman JB, Slavin AJ, Silva R et al (2000) Leukocyte recruitment during onset of experimental allergic encephalomyelitis is CCR1 dependent. Eur J Immunol 30:2372–2377
9. Fife BT, Huffnagle GB, Kuziel WA et al (2000) CC chemokine receptor 2 is critical for induction of experimental autoimmune encephalomyelitis. J Exp Med 192:899–906
10. Forde EA, Dogan RN, Karpus WJ (2011) CCR4 contributes to the pathogenesis of experimental autoimmune encephalomyelitis by regulating inflammatory macrophage function. J Neuroimmunol 236:17–26
11. Carlson T, Kroenke M, Rao P et al (2008) The Th17-ELR+CXC chemokine pathway is essential for the development of central nervous system autoimmune disease. J Exp Med 205:811–823
12. Karpus WJ, Lukacs NW, McRae BL et al (1995) An important role for the chemokine macrophage inflammatory protein-1α in the pathogenesis of the T cell-mediated autoimmune disease, experimental autoimmune encephalomyelitis. J Immunol 155:5003–5010
13. Huang DR, Wang J, Kivisakk P et al (2001) Absence of monocyte chemoattractant protein 1 in mice leads to decreased local macrophage recruitment and antigen-specific T helper cell type 1 immune response in experimental autoimmune encephalomyelitis. J Exp Med 193:713–726
14. Dogan RN, Long N, Forde E et al (2011) CCL22 regulates experimental autoimmune encephalomyelitis by controlling inflammatory macrophage accumulation and effector function. J Leukoc Biol 89:93–104
15. Elhofy A, Depaolo RW, Lira SA et al (2009) Mice deficient for CCR6 fail to control chronic

experimental autoimmune encephalomyelitis. J Neuroimmunol 213:91–99
16. Shen FW, Saga Y, Litman G et al (1985) Cloning of Ly-5 cDNA. Proc Natl Acad Sci USA 82:7360–7363
17. Chvatchko Y, Hoogewerf AJ, Meyer A et al (2000) A key role for CC chemokine receptor 4 in lipopolysaccharide-induced endotoxic shock. J Exp Med 191:1755–1764
18. Roederer M (2001) Spectral compensation for flow cytometry: visualization artifacts, limitations, and caveats. Cytometry 45:194–205
19. Miller SD, Karpus WJ, Davidson TS (2010) Experimental autoimmune encephalomyelitis in the mouse. Curr Protoc Immunol Chapter 15, Unit 15, 11
20. Pope JG, Karpus WJ, VanderLugt C et al (1996) Flow cytometric and functional analyses of central nervous system-infiltrating cells in SJL/J mice with Theiler's virus-induced demyelinating disease: evidence for a CD4+ T cell-mediated pathology. J Immunol 156:4050–4058
21. Stewart CC, Stewart SJ (2001) Titering antibodies. In: Paul Robinson J (ed) Current John Wiley & Sons, protocols in cytometry, pp 4.1.1–4.1.3
22. Geoffrey WO (2011) Chapter 21: Recent advances in flow cytometric cell sorting. In: Darzynkiewicz Z, Holden E, Orfao A, Telford W, Wlodkowic D (eds) Methods in cell biology, vol 102. Academic press headquarters is Waltham, Massachusetts, New York, pp 533–556

Chapter 11

Visualizing Chemokine-Dependent T Cell Activation and Migration in Response to Central Nervous System Infection

Monica J. Carson and Emma H. Wilson

Abstract

In response to central nervous system (CNS) injury and infection, astrocytes, neurons, and CNS vasculature express several chemokines, including CCL21. Quantitative polymerase chain reaction (qPCR), western blot, and immunohistochemical methods can quantify mRNA and protein expression. However, these methods do not quantify chemokine bioavailability and bioactivity, variables modified by many environmental factors including composition of the extracellular matrix (ECM). Here we illustrate how two-photon microscopy and carboxyfluorescein succinimidyl ester (CFSE or CFDA SE) labeling of T cells coupled with flow cytometry can be used as tools to assess chemokine-mediated regulation of T cell proliferation, activation, and migration.

Key words CFSE, Multiphoton microscopy, Cell migration, Cell proliferation, Second harmonic structures, Flow cytometry

1 Introduction

Chemokine expression is a critical aspect of the central nervous system (CNS) response to injury and infection [1–4]. Each chemokine serves as a recruiting, organizing, and activating agent for selected populations of immune cells expressing the appropriate receptors. However, distinguishing between immune responses for which a specific chemokine is necessary versus sufficient can be difficult due to the multiple inflammatory and cross-regulatory events triggered by injury and infection. An additional confound is the complexity of the multiple ligand–receptor relationships between chemokines and their receptors. For example, CCR7 is the high-affinity receptor for both CCL19 and CCL21, but each ligand triggers distinct biological effects [1–3]. Notably, while CCL19 is much more effective in supporting T cell activation and survival in vitro, CCR7-expressing T cells preferentially migrate toward CCL21 in a competing gradient in vitro chemotaxis assay.

Astrid E. Cardona and Eroboghene E. Ubogu (eds.), *Chemokines: Methods and Protocols*, Methods in Molecular Biology, vol. 1013, DOI 10.1007/978-1-62703-426-5_11, © Springer Science+Business Media New York 2013

In response to CNS *Toxoplasma gondii* (*T. gondii*), astrocytes express high levels of CCL21 while CCR7-expressing T cells infiltrate the CNS and home to foci of infection [3, 4]. Because CCL21 is a potent T cell chemoattractant, it is tempting to presume that CCL21 is sufficient to recruit T cells from the blood circulation and direct them to sites of infection, extrapolating correlative observations to causation. However, in support of this hypothesis, CCL21 expression was found to be both necessary and sufficient to recruit $CD4^+$ T cells into the pancreas in a murine model of autoimmune diabetes [5]. Therefore, to directly test the hypothesis that CCL21 performs the same role in the CNS, we generated transgenic mice in which astrocytes constitutively express slightly higher levels of CCL21 than observed in *T. gondii* infection [3].

Surprisingly, not only was there an absence of microglial activation, but also no T cells were found in the CCL21-expressing CNS. These data suggested that CNS-expressed CCL21 (a) was not detected by T cells located outside the CNS, (b) was not bioactive, or (c) was insufficient by itself to induce CCR7-expressing T cell infiltration into the CNS. Using CCL21 deficient and overexpression mouse models, we found that CCL21 while bioactive was insufficient to support T cell migration into the CNS. Conversely, we also found that CCL21 expression was required for $CD4^+$ T cell migration from sites of extravasation to sites of *T. gondii* infection within the CNS [3]. In the following sections, we present two methods: (1) carboxyfluorescein succinimidyl ester (CFSE) labeling of T cells that allows simultaneous quantitation of cell proliferation and activation while brightly tagging cells for microscopic evaluation and (2) multiphoton-based microscopic imaging of cell migration within living tissue [5–10].

1.1 Using CFSE to Quantify Bioavailability of CNS-Expressed CCL21

Bioavailable CCL21 drives antigen-independent $CD4^+$ T cell proliferation accompanied by upregulation of CD69 (facilitates T cell receptor responses), as well as adhesion molecules associated with extravasation and infiltration into the CNS [5]. For decades one of the most common methods for monitoring cell proliferation is based on quantifying radiolabeled nucleotide incorporation into the DNA of dividing cells [7–10]. However, this method is most appropriate in demonstrating proliferation at the level of an entire cell population. It does not provide information on the number of cell divisions completed per cell and the percentage of cells proliferating nor can it be readily combined with flow cytometric assays to quantify the expression of immune activation markers.

Alternatively, modified nucleotides such as bromodeoxyuridine (BrdU) can be fluorescently labeled so that proliferating cells can be identified in total cell populations by flow cytometric analysis or by fluorescent microscopy. This approach also allows simultaneous quantitation of markers of cell activation by flow cytometry, as well as provides spatial information regarding sites of proliferation when analyzing tissue sections. However, these approaches do not readily discern the number of cell division cycles by each fluorescent cell. In addition, detection in histologic section is difficult and may require signal amplification by a variety of methods. All of these limitations can be overcome by using CFSE [7–10].

CFSE is a nonfluorescent, cell-permeable fluorescein derivative. CFSE only becomes fluorescent after it diffuses into a cell's cytoplasm and its acetate groups are cleaved by cellular esterases [7–10]. The succinimidyl ester then binds free amine groups of other molecules within the cell resulting in long-lived fluorescent labeling of treated cells. These CFSE-modified molecules referred to as adducts are retained within the cell and distribute equally in daughter cells following cell division. Thus, with each cell division, CFSE fluorescence decreases by half. Enumeration of cell division cycles completed per cell is readily analyzed by either flow cytometry of cell suspensions or confocal microscopy of living or fixed tissues using detectors and filters appropriate for fluorescein (FITC)-range fluorescent emissions. In our hands, 6–8 cell divisions are easily detected relative to fluorescence levels of unlabeled cell using standard flow cytometry (Fig. 1). In addition, this method is readily combined with other standard flow cytometry-dependent assays of cell activation/differentiation (using

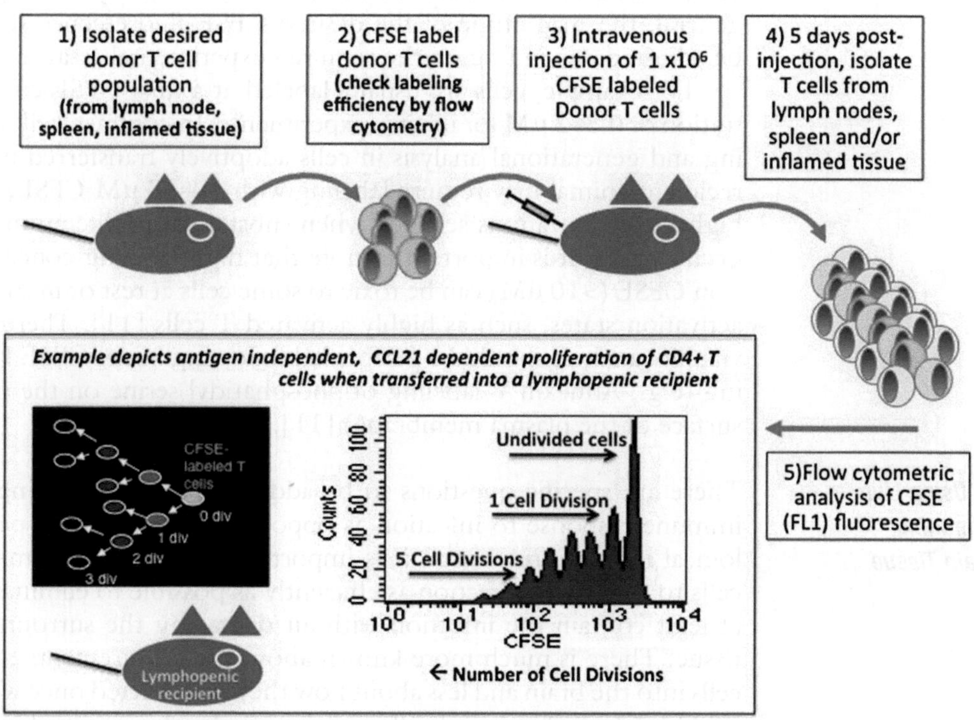

Fig. 1 CFSE labeling to quantify CCL21-dependent CD4⁺ T cell proliferation

fluorescently conjugated antibodies or other dyes detected outside the FITC range of fluorescence emission). However, a few essential caveats need to be considered when designing experiments using CFSE-labeled cells.

First, although CFSE generates stable fluoresceinated adducts, the dye fades with time [7–10]. Therefore, to deduce fluorescence levels that corresponds to undivided cells, it is essential to fix (e.g., 1–2 % paraformaldehyde [PFA] in 1× phosphate-buffered saline [PBS]) and save an aliquot of the CFSE-labeled cells at 4 °C when initially labeling. Furthermore, autofluorescence varies as a function of cell type, activation state, or viability, or a combination thereof. Therefore, it is also essential to determine the autofluorescence of unlabeled cells at equivalent activation states in order to define the range that CFSE can be used as a marker of proliferation.

Second, the absolute fluorescence intensity level is directly proportional to the concentration of CFSE dye used to label a cell population [7–10]. During flow cytometry CFSE fluorescence is detected in the FL-1 channel. However, there is also substantial spread of fluorescein emission into the FL-2 channel that may potentially interfere with simultaneous detection of phycoerythrin (PE)-conjugated antibodies or other labels emitting in FL2 channel. This problem can be overcome in part by either flow cytometer hardware- or analytical software-based compensation. An alternative approach is to alter the initial labeling conditions to decrease the overall level of cellular fluorescence when initially labeling. In general, the magnitude of the desired CFSE fluorescence should be adjusted depending on the required experimental assay.

For example, cells are usually labeled at a final CFSE concentration of 0.5–2 μM for in vitro experiments. In contrast, cell tracking and generational analysis in cells adoptively transferred into a recipient animal may require labeling with 10–50 μM CFSE, with higher concentrations selected when substantial proliferation may occur. Lastly, it is important to note that high-labeling-concentration CFSE (>10 μM) can be toxic to some cells at rest or in specific activation states, such as highly activated T cells [11]. Therefore, viability analysis should be performed following initial CFSE labeling (e.g., Annexin V labeling of phosphatidyl serine on the outer surface of the plasma membrane) [11].

1.2 Using Live Cell Imaging in Brain Tissue

There are specific questions to be addressed when analyzing the immune response to infection as opposed to generating a pathological immune response—it is important to guide inflammatory cells to the site of infection as efficiently as possible to eliminate or at least contain the infection without damaging the surrounding tissue. There is much more known about the requirements to get cells into the brain and less about how they are directed once within it [11]. Chemokines and inflammatory molecules involved in this process are therefore of key interest.

Although several methods can be employed to study the infected brain's immune response, including immunohistochemistry and flow cytometry, the migration and behavior of cells within the brain can only be studied using multiphoton (MP) microscopy. There are many excellent reviews on the use of this type of microscopy [12, 13]. However there are three major aspects that make MP-imaging the only solution to investigating cell behavior within tissues: (1) It uses an infrared pulsing laser that has a longer wavelength allowing deeper tissue penetration; (2) a longer wavelength results in lower energy-lessening damage to the tissue, allowing longer imaging time; and (3) excitation only occurs at the focal plane as two photons are required to provide sufficient energy to excite the fluorophore. This prevents excitation above and below the focal plane allowing clear signal-to-noise ratio. These features of MP microscopy allow fluorescently labeled cells to be imaged at depths of up to ~300 μm for hours at a time.

The following method describes the procedure for tracking green fluorescent protein (GFP)-expressing T cells in an excised brain. Removal of the tissue is possibly the most straightforward procedure; however there are some caveats that can be overcome with choosing alternative procedures. Firstly, although the brain is kept alive in temperature- and oxygen-controlled environments, the tissue does not have a blood supply; therefore this procedure should not be used if the question being asked involves extravasation. Instead the vasculature can be kept intact by removing the skull and imaging the brain within an alive, terminally anesthetized mouse. Secondly, this is a terminal procedure! Therefore it is not possible to ask questions that may require imaging the same mouse over several days or weeks. Instead the skull can be left in place but thinned to allow a window for imaging and a specific area that can be returned to indefinitely. This method requires a lot more skill, technique, and equipment; however as it also leaves the surface of the brain intact, it avoids the activation of glial cells and, as a result, is essentially the gold standard for live cell imaging.

The advantages of removing the entire brain include being able to use second harmonic generation (SHG) to image structures in the brain. This is a phenomenon specific to MP microscopy and is the result of photons being combined on non-centrosymmetric structures and being scattered back at exactly half the wavelength and twice the energy. This has been observed with collagen structures in the lymph node, tumor environment, and granulomas [12, 13]. We have also observed inflammation-induced structures in the CNS that associate with migrating lymphocytes [6]. The intact skull would result in such a large SHG signal and prevent the collection of any signal from within the tissue.

Furthermore, although the lack of a blood supply hinders certain analyses, it reduces z-movement to a minimum. When imaging for long periods, slight changes in the z-position can result in

Table 1
In vivo measurements using multiphoton microscopy

In vivo measurements	Definition	Interpretation
Velocity	Speed of cells	Cell velocity decreases in the presence of antigen; velocity <2 μm/min is considered stationary
Distance	Total distance travelled from point A to B	Distance covered may include repetitive behavior and is not the best indicator of how directed a cell is
Displacement	Actual distance travelled from point A to B	Fast highly directed cells are expected to have greater displacement indexes
Meandering index	Measurement of directionality	Distance travelled/displacement: The closer the meandering index is to 1.0 the more directed the cell is travelling from point A to B
Clustering	Group of cells with distinct behavior	Clusters of slow or stationary cells implies antigen-specific responses
SHG signal	Volume of reticular fibers	Inflammation-induced reticular fiber migration of cells is associated with these fibers

the loss of the original imaging area and prohibit cell tracking. Lastly, the advantage of excising the brain allows the whole tissue to be imaged without confinement to a portion of the brain where the skull has been thinned. This is particularly useful when imaging infection as, unlike injury models, the infected area is not always known, so the ability to scan a relatively large area is beneficial.

The type of data obtained from these experiments will consist of the velocity and directionality of cells. Table 1 provides an overview of the types of analysis conducted and what the data may mean. For example, a T cell travelling at a speed less than 2 μm/min is considered stationary and likely undergoes an antigen-dependent function, e.g., interaction with an infected cell. Also recent studies have demonstrated that T cells search the infected brain tissue in a traditional generalized Lévy walk dependent on chemokine signaling. This type of behavior maximizes the chances of the T cell finding a rare event—an infected cell. Thus, analysis of cell behavior is becoming more sophisticated but care should be taken to use a large data sample (number of cells and length of time) if generalized statements about whole cell population behavior are required.

Lastly, multiple cell labeling can be hugely advantageous to deciphering cell behavior. Examples include using mice with endogenously labeled microglia or astrocytes in the brain to allow cell–cell interactions to be measured or if the source of infection is also fluorescently labeled. However, the ability to image multiple fluorochromes will depend on the number of detectors available in the microscope setup and the ability to clearly distinguish overlapping emission signals. As this type of imaging becomes more widespread with the continued development of microscopes, it is likely that MP-imaging of multiple cells and their interactions will become possible.

2 Materials

2.1 CFSE Labeling and Detection

1. Purified cell population (*see* **Note 1**).
2. CFSE stock solution (10–50 mM) in dimethyl sulfoxide (DMSO) (*see* **Note 2**).
3. 1× phosphate-buffered saline (PBS) or Hank's balanced salt solution (HBSS) plus 1 % bovine serum albumin (BSA) for labeling reaction.
4. Complete tissue culture medium, RPMI containing 10 % fetal bovine serum for washes and final resuspension (*see* **Note 3**).
5. Flow cytometer or confocal microscope equipped with lasers, detectors, and filters appropriate for fluorescein isothiocyanate (FITC) fluorescence.
6. Software to quantify fluorescence. CellQuest™ and FlowJo can be used for acquired flow cytometric data (*see* **Note 4**) and Volocity (Perkin Elmer) can be used for acquired confocal/multiphoton data.

2.2 Live Cell Tracking in Brain Tissue

1. Mouse with appropriate fluorophore-labeled cells within the brain (*see* **Notes 5** and **6**).
2. Dissection equipment and complete tissue culture medium. An MP microscope, i.e., one that has an infrared pulsed laser—this protocol uses a Zeiss LSM710 NLO microscope with Coherent laser with two non-descanning detectors (NDD) to detect SHG (425 nm) and GFP (530 nm) (*see* **Note 7**).
3. Live imaging setup attached to the microscope, i.e., temperature-controlled chamber and flow of oxygenated medium; *see* **Note 8**.
4. Analysis software, e.g., Volocity.

3 Methods

3.1 CFSE Labeling and Detection

1. Isolate T cells from lymph nodes, spleen, or specific tissues of interest. If activation, differentiation, or viability assays are required prior to CFSE labeling, cells can be treated with relevant antibodies or other labeling reagents at this point.

2. Suspend the cells at a concentration of 1×10^7 cells/ml or less in 1× PBS or HBSS containing 0.1 % BSA. Total reaction volumes should not exceed 4 ml in a conical 15 ml tube, so prepare cell suspensions at no greater than 2 ml each.
3. Prepare a solution of CFSE from DMSO stock containing 0.1 % BSA in 1× PBS at twice the final labeling concentration. For example, if you are labeling at 5 μM, prepare a 10 μM solution. Prepare the volume of CFSE equal to your cell volume above (no more than 2 ml per labeling reaction).
4. Add an equal volume of CFSE solution to your cell suspension. Mix gently and incubate for up to 10 min at 37 °C (*see* **Note 4**).
5. Immediately fill the labeling tube to the top with complete tissue culture medium and centrifuge for 7 min, 600 × *g*. Wash the cells three times with 3–5 ml of tissue culture medium at room temperature or 4 °C (*see* **Note 9**).
6. Resuspend at the concentration/volume appropriate for injection (50×10^6 cells/ml) or cell culture (2×10^6 cells/ml).
7. Immediately fix an aliquot of the labeled cells in 1–2 % paraformaldehyde solution (in 1× PBS) to provide a CFSE fluorescence standard in nondividing cells.
8. Perform flow cytometry on a small aliquot of labeled cells (50,000–100,000 cells) to determine the range of CFSE fluorescence intensity per cell and general cell viability with a flow cytometric instrument capable of detecting FL1/FITC-type fluorescence.
9. Cells are now ready for transfer into cultured tissues, living recipient mice, or for in vitro use in activation and proliferation assays. Following completion of the experiment, harvest the cells of interest from culture or the lymph nodes, spleen, or other tissues of the recipient animal. Analyze both the recovered cells and the stored fixed cell samples (*see* **step** 7 above) obtained at the time of initial labeling at the same time (by flow cytometry or microscopy).

3.2 Live Imaging in Brain Tissue

1. Excise the mouse brain and place in a temperature-controlled chamber.
2. Set the objective (*see* **Note 10**).
3. Look through the eyepiece using normal argon lamp fluorescence and focus. Superficial brain autofluorescing blood vessels should be clearly seen, providing targets to facilitate focus.
4. Find your field of view—this will depend on your experimental question; however this should include an area of fluorescent cells. At this point, it is often possible to detect highly migratory cells through the eyepiece alone.

5. Switch to the two-photon setup by turning down the argon lamp and switching to NDD (*see* **Note 11**).
6. Confirm signal by live tissue scanning (*see* **Note 12**).
7. Set up a z-series. The z-series should be thick enough to capture a significant portion of tissue allowing the SHG structure to be noted and so that migrating cells can be followed for a significant length of time. We suggest a z-series of six steps separated by 6 μm giving a total z-stack of 30 μm. Increasing the step size may result in a migrating cell being lost between slices; increasing the total height of the stack will increase the length of time it takes to image—crucial when combining with a time-series.
8. Set up a time-series. The time-series needs to be long enough so that the migration and behavior of cells can be analyzed. The time-series is based on how many consecutive repeats of the z-stack are to be imaged. Therefore the length of the movie generated will depend on the time it takes to image one z-stack. A minimum of 20 time points should be imaged to generate enough data to make conclusions about velocity and direction of migrating cells.
9. Run the time-series (*see* **Note 13**).
10. Analysis of cell migration requires specialized software. Few programs allow tracking of cells in four dimensions (x, y, z, and time). We have successfully used Volocity that allows manual and automatic cell tracking (*see* **Note 14**).
11. Using Volocity, import the new image sequence from acquisition files.
12. Modify auto contrast and color as desired (e.g., green for GFP).
13. Go to "measurements view" and "measure all time points."
14. Find objects—this can be done in several ways depending on the level of background, number of cells, and measurements wanted. The following is an example:
15. Find by % intensity (cut off lower and upper 5 % and look at histogram).
16. Then exclude by size (most cells are >200 μm^3).
17. Track items—here essentially the formula for finding the next position of the cell defined—shortest path is normally the best using <10 μm distance.
18. Once tracks are obtained, highlight the track objects, go back to the image view, and play the movie checking that the tracks make sense.
19. Save as a measurement item. Each tracked object will have a mass of data values presented as a table including average

velocity, meandering index, and displacement. The resulting data can be copied and pasted into an external software spreadsheet for subsequent analyses.

4 Notes

1. CFSE labeling is most appropriate for cells that are fairly homogeneous in cell shape such as lymphocytes. As a consequence, the initial labeling of a homogeneous cell population will have a fairly uniform level of fluorescence per cell. When graphed, the entire population of labeled cells has a narrow range of fluorescence intensity (see narrow peak of labeled cells in Fig. 1). If cells are initially heterogeneous in morphology and size (such as observed with many macrophage-like cells or cultured cells), then the range of fluorescence intensity per cell will be fairly broad and thus interfere with definitive detection of proliferating cells with half of the fluorescence of undivided cells.
2. Prepare the CFSE at a stock concentration 1,000-fold higher than the final usage concentration in DMSO. Aliquot CFSE solution into single-usage vials and store under desiccating conditions at −20 °C. CFSE will hydrolyze quickly at room temperature with water present, and much more slowly at −20 °C under desiccating conditions. Aliquoted stocks should be used for no more than 2 months.
3. It is essential to maintain cells in single-cell suspension and completely mix the cells with the CFSE dye. The 1:1 ratio of cells to dye facilitates even labeling of cells such that a narrow range of fluorescence per cell is generated.
4. Software programs such as ModFit can be used to estimate cell divisions in data derived from cells with an initial broad range of labeling intensities.
5. Many strains of mice are available that have endogenously controlled fluorescence. A list available from The Jackson Laboratory can be found at the following Web site: http://jaxmice.jax.org/research/fluorescent_proteins_or_lacZ.html.
6. Imaging and analysis are very time consuming (and expensive if paying for microscope time); therefore it is worth checking for fluorescent cells prior to imaging using another method (e.g., flow cytometry or immunohistochemistry) especially if the cells are being transferred or are not endogenous to the mouse. It is also useful to optimize the number of cells in the brain, e.g., too few migrating T cells will require several brains to generate significant data on cell behavior. Too many migrating T cells make analysis extremely difficult due to overlapping cell tracks.

7. The fluorophore needs to be excitable by the two-photon laser and detectable by your microscope. Thus GFP with an excitation range of between 400 and 530 nm requires a two-photon laser wavelength of between 800 and 1,060 nm. At present, few two-photon lasers reach beyond 1,100 nm and the majority do not function above 950 nm.
8. Anything other than excised tissue will require an upright microscope or an inverter. The temperature-controlled chamber is often set up so that a 60 mm Petri dish fitted with a 1.5 mm diameter cover glass floor fits in the center. The thickness of the cover glass is crucial.
9. To reduce the amount of unbound CFSE in cells, incubate the cells at 37 °C for 5 min between the second and third washes; this facilitates the release of free non-reacted CFSE by diffusion out of the cells for removal in the final wash.
10. If using an upright setup, a dipping objective is required with appropriate working distance; otherwise a 40× water objective is needed to provide the best field of view in our experience. When using water objectives, make sure that the water is added to the objective prior to mounting the sample as it can be challenging to add afterwards. Avoid using medium containing pH indicators as this can generate autofluorescence.
11. NDDs are highly sensitive allowing maximum sample signal detection. However, it also means they can pick up ambient light, including light from the computer monitor. Minimize sources of light in the room and cover detectors and microscope with blackout curtains.
12. At this point the SHG signal can be seen as long fibers or waves especially in the area of blood vessels—presumed to be collagen. GFP-positive cells will be clearly observed alone in the single channel—small autofluorescing green blobs are often seen but show up in all channels and are much smaller than T cells (Fig. 2).
13. When running the time-series, it may be apparent that there is z-movement. This may happen especially in thick tissues such as the brain as it settles or if using a live mouse setup due to respiratory movements, etc. Pick a landmark in a *z*-plane and watch for changes or appearance in a different plane. If this occurs, then the movie should be quickly stopped. Reset the *z* start and stop positions to account for the movement and restart the movie *s*. During data analysis, files can be combined to join the break in the movie sequence.
14. When using the manual tracking feature, the cell's center needs to be indicated at every time point using *x*-, *y*-, and *z*-planes. This introduces user error or bias but may be required sometimes when there are many overlapping tracks. When using the

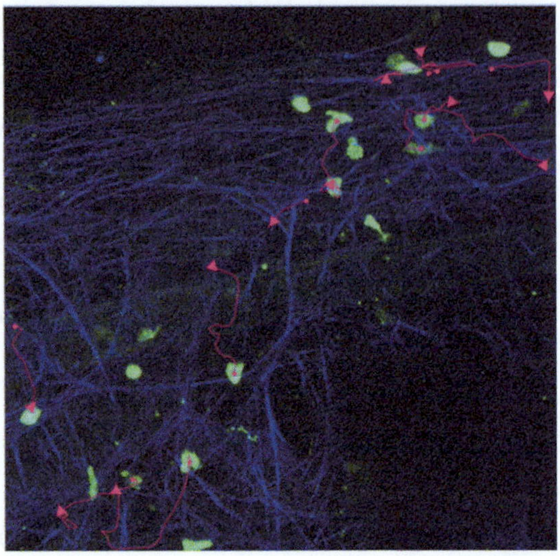

Fig. 2 Snapshot image showing SHG signal (*blue*), eGFP (*green*), and tracks of migrating T cells (*pink*)

automated tracking settings, the parameters have to be defined carefully to find only objects that are cells. Following tracking, observe selected tracks to ensure that they adequately account for the expected cell behavior.

Acknowledgements

Assay development and data generation are supported by a University of California—Riverside Chancellor's strategic initiative and National Institutes of Health grants NS071160A and NS072298A. The authors acknowledge Corinne Ploix for contributions toward standardization of flow cytometric protocols and Sean Wilson and Monica Rubalcava (Loma Linda Univerisity School of Medicine) for contributions toward optimization of multiphoton microscopy protocols.

References

1. Lalor SJ, Segal BM (2010) Lymphoid chemokines in the CNS. J Neuroimmunol 224:56–61
2. Engelhardt B (2006) Molecular mechanisms involved in T cell migration across the blood brain barrier. J Neural Transm 113:477–485
3. Ploix CC, Noor S, Crane J, Masek K, Carter W, Lo DD, Wilson EH, Carson MJ (2011) CNS-derived CCL21 is both sufficient to drice homeostatic CD4+ T cell proliferation and necessary for efficient CD4+ T cell migration into the CNS parenchyma following Toxoplasma gondii infection. Brain Behav Immun 25:883–896
4. Noor S, Habashy AS, Nance JP, Clark RT, Nemati K, Carson MJ, Wilson EH (2010) CCR7-dependent immunity during acute

Toxoplasma gondii infection. Infect Immun 78:2257–2263

5. Ploix C, Lo D, Carson MJ (2001) A ligand for the chemokine receptor CCR7 can influence the homeostatic proliferation of CD4+ T cells and progression of autoimmunity. J Immunol 167:6724–6730
6. Wilson EH, Harris TH, Mrass P, John B, Tait ED, Wu GF, Pepper M, Wherry EJ, Dzierzinski F, Roos D, Haydon PG, Laufer TM, Weninger W, Hunter CA (2009) Behavior of parasite-specific effector CD8+ T cells in the brain and visualization of a kinesis-associated system of reticular fibers. Immunity 30:300–311
7. Parish CR (1999) Fluorescent dyes for lymphocyte migration and proliferation studies. Immunol Cell Biol 77:499–508
8. Lyons AB (2000) Analysing cell division in vivo and in vitro using flow cytometric measurement of CFSE dye dilution. J Immunol Methods 243:147–154
9. Lyons AB, Hasbold J, Hodgkin PD (2001) Flow cytometric analysis of cell division history using dilution of carboxyfluoresceindiacetate succinimidyl ester, a stably integrated fluorescent probe. Methods Cell Biol 63:375–398
10. Li CR, Santoso S, Lo DD (2007) Quantitative analysis of T cell homeostatic proliferation. Cell Immunol 250:40–54
11. Wilson EH, Weninger W, Hunter CA (2010) Trafficking of immune cells in the central nervous system. J Clin Invest 120:1368–1379
12. Textor J, Peixoto A, Henrickson SE, Sinn M, von Andrian UH, Westermann J (2011) Defining the quantitative limits of intravital two-photon lymphocyte tracking. Proc Natl Acad Sci USA 108:12401–12406
13. Bajénoff M, Germain RN (2007) Seeing is believing: a focus on the contribution of microscopic imaging to our understanding of immune system function. Eur J Immunol 37(S1): S18–S33

Chapter 12

Epigenetic Regulation of Chemokine/Chemokine Receptor Expression

Anne-Marie Baird, Kenneth J. O'Byrne, and Steven G. Gray

Abstract

Epigenetic regulation of gene expression is an important event for normal cellular homeostasis. Gene expression may be "switched" on or "turned" off via epigenetic means through adjustments in DNA architecture. These structural alterations result from changes to the DNA methylation status in addition to histone posttranslational modifications such as acetylation and methylation. Drugs which can alter the status of these epigenetic markers are currently undergoing clinical trials in a wide variety of diseases, including cancer.

We illustrate the treatment of cell lines with histone deacetylase (HDi) and DNA methyltransferase inhibitors and the subsequent RNA isolation and reverse transcriptase polymerase chain reaction for several members of the CXC (ELR+) chemokine family. In addition we describe a chromatin immunoprecipitation assay to determine the association between chromatin transcription markers and DNA following pretreatment of cell cultures with an HDi, Trichostatin A (TSA). This assay allows us to determine whether treatment with TSA dynamically remodels the promoter region of our selected genes, as judged by the differences in the PCR product between our treated and untreated samples.

Key words CXC chemokines, Histone deacetylase, DNA methyltransferase, Trichostatin A, RNA, RT-PCR, Chromatin immunoprecipitation

1 Introduction

It was in 1942 that Waddington coined the term epigenetics [1]. It is defined as the study of heritable changes in gene expression without a change in DNA sequence. DNA conformation dictates the expression patterns of genes, as it determines when and where along the DNA sequence transcription factors can bind to transcribe genes. Epigenetic changes through alterations in the DNA methylation status and histone posttranslational modifications can adjust this access, therefore regulating gene expression. Adverse epigenetic events occur frequently in cancer and may assist in the activation of proto-oncogenes or the silencing of tumor suppressor genes, or both [2].

Astrid E. Cardona and Eroboghene E. Ubogu (eds.), *Chemokines: Methods and Protocols*, Methods in Molecular Biology, vol. 1013, DOI 10.1007/978-1-62703-426-5_12, © Springer Science+Business Media New York 2013

Chromatin immunoprecipitation (ChIP) is a method used to determine the interactions between proteins and DNA that was first described nearly 30 years ago. The technique is continuously being refined and improved [3, 4]. The ChIP assay was first developed by Gilmour and Lis in the mid 1980s when they examined the association of RNA polymerase II initially to bacterial genes [5] and subsequently *Drosophila melanogaster* heat-shock genes [6, 7]. The method consisted of cross-linking DNA and proteins with ultraviolet (UV) light, isolating specific DNA complexes by immunoprecipitation and characterizing these complexes by hybridization [5–7].

Although DNA is still cross-linked using UV light, formaldehyde is now more commonly used as the cross-linking agent. One of the first studies to identify the use of formaldehyde as a cross-linking agent was performed as early as 1978 [8]; however it was the Varshavsky Group in Massachusetts Institute of Technology that described the use of formaldehyde in ChIP (referred to as X-ChIP) [9], employing it to study the association between RNA polymerase II and histone H4 with *D. melanogaster* heat-shock genes. Cross-links formed were reversed by heat following the immunoprecipitation steps [9]. The ChIP assay has also been modified to use chromatin prepared without a cross-linking step [10] and DNA is fragmented using micrococcal nuclease digestion. This is referred to as native ChIP or N-ChIP. N-ChIP allows for the identification of tightly associated chromatin proteins, while X-ChIP provides a larger pool of associated proteins [11].

The applications of the ChIP assay have been greatly enhanced through the production of specific antibodies directed towards histone markers and modifications and nonhistone proteins. Significant advances have been made since the first antibody used against *N*-acetyl-lysine in 1988 [12]. Other major improvements have come through the adaptation of ChIP for use in mammalian systems with DNA cross-linking to proteins [13, 14]. Samples derived from a small number of cells such as stem cell populations can now also be used in ChIP studies, for example using carrier ChIP (C-ChIP), whereby carrier chromatin from an evolutionary distant species (such as *Drosophila*) is mixed with the mammalian cells [3, 15]. However, a fast one-day micro ChIP assay has been recently described without using carrier chromatin [16].

The endpoint assessment of the ChIP assay has also changed from hybridization assays [5] through to polymerase chain reaction (PCR)-based detection methods such as semiquantitative PCR [17] and quantitative PCR (qPCR) [18]. Currently, there are many modified ChIP assay protocols with a wide variety of endpoint readouts, for example (a) ChIP-on-chip, which combines ChIP with micro-array-based technology which can encompass whole genome analysis and (b) sequence-based methods with ChIP such as ChIP-SAGE (serial analysis of gene expression), ChIP-PET (paired end ditags), and ChIP-seq (direct sequencing) (reviewed in ref. 19).

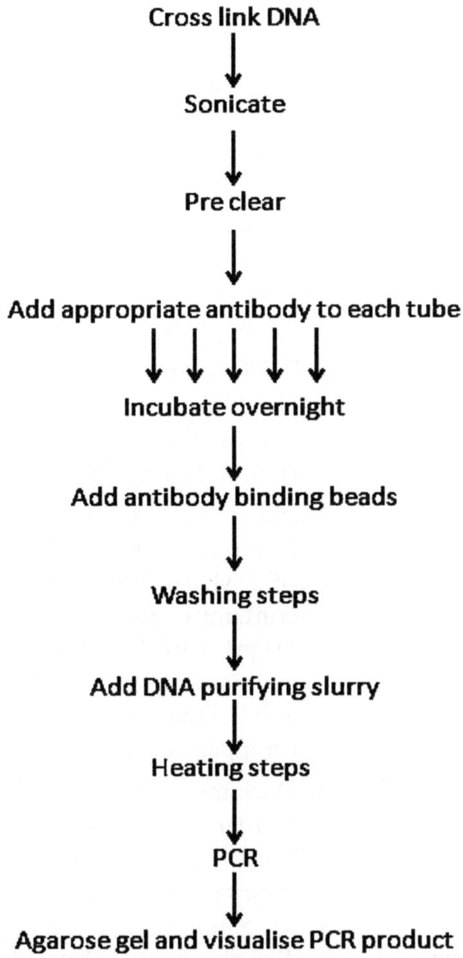

Fig. 1 Overview of the ChIP protocol. The flow diagram outlines the ChIP assay described in detail in Subheadings 3.4–3.7

Here we describe the response of CXC (ELR^+) genes to epigenetic treatment, through RNA isolation and reverse transcriptase polymerase chain reaction (RT-PCR). We also illustrate an X-ChIP assay to determine the effect of a histone deacetylase inhibition (HDi) on the promoter region of our genes of interest using PCR with primers designed to the 5′ untranslated regions (UTRs) contained within their nucleotide sequences. A flowchart of the described ChIP method is outlined in Fig. 1.

2 Materials

The reagents and chemicals used should be of analytical grade and stored in accordance with manufacturers' instructions. Buffers and solutions should be prepared using distilled water and stored at room temperature unless stated otherwise. All local regulations in relation to the handling and disposal of reagents and chemicals must be followed.

2.1 Epigenetic Drug Treatments

Epigenetic drugs have varying solubility requirements; therefore data sheets should be consulted before use (*see* **Note 1**). The solvent used will determine the vehicle control (*see* **Note 2**). Stocks of each drug must be made aseptically and aliquoted into smaller volumes to prevent freeze–thaw damage (where applicable) and minimize the risk of contamination.

1. Trichostatin A (TSA). Dissolve TSA in dimethyl sulfoxide (DMSO) to a concentration of 250 μg/mL and store at −20 °C. Treat cell cultures (*see* **Notes 3** and **4**) with a final concentration of 250 ng/mL for a period of 24 h (*see* **Note 5**).
2. Phenylbutyrate (Tributyrate™) [PB] (A gift from Triple Crown America, Perkasie, PA, USA). Make a working stock solution of 2 M PB: Weigh 3.724 g PB and transfer to a 50 mL tube containing 5 mL phosphate-buffered saline (PBS). Make up to 10 mL with PBS and sterile filter (0.8 μM filter) in a Class IIB laminar airflow unit. Prepare fresh before each use and store at 4 °C. Treat cell cultures with a final concentration of 10 mM for a period of 24 h.
3. Decitabine (5-aza-2′-deoxycytidine) [DAC]. Dissolve in 100 % methanol to a concentration of 4.38 mM and store at −20 °C. Treat cell cultures with a final concentration of 1 μM for 48 h. Replace DAC and cell culture medium every 24 h (*see* **Note 6**).

2.2 RNA Isolation

1. TRI reagent® (Molecular Research Center, Cincinnati, OH, USA). We consistently obtain a large concentration of pure RNA using this particular reagent. It also facilitates DNA and protein isolation from a single sample, saving on time and expense associated with using several reagents.
2. 1-bromo-3-chloro-propane (BCP).
3. Isopropanol (2-isopropanol).
4. Diethylpyrocarbonate (DEPC)-treated water.
5. Ethanol (EtOH) wash buffer: 70 % solution in water.

2.3 RNA-Qualified DNAse Treatment

1. RNA-qualified (RQ1) RNase-free DNase.
2. Phenol:chloroform:isoamyl alcohol in ratio 25:24:1. Saturated with 10 mM Tris, pH 8.0, 1 mM EDTA (*see* **Note 7**).
3. Ammonium acetate: 5 M solution in water. Weigh 19.27 g ammonium acetate and initially dissolve with a small amount of water (*see* **Note 8**). Add water to make up to 50 mL.
4. Polyacryl carrier (Molecular Research Center, Cincinnati, OH, USA) (*see* **Note 9**).
5. 96% Ethanol.

2.4 Complementary DNA Synthesis

1. Oligo(dT)$_{20}$ (*see* **Note 10**).
2. 10 mM dNTP stock (*see* **Note 11**).
3. Ribonuclease inhibitor.
4. 5× First-strand (FS) buffer.
5. Reverse transcriptase enzyme.

2.5 ChIP Assay

1. High-quality 37 % formaldehyde solution.
2. Sodium dodecyl sulfate (SDS) buffer: 10 % solution in water (*see* **Note 12**).
3. Tris buffer: 100 mM solution in water. Weigh 6.056 g Tris base and transfer to a graduated cylinder containing approximately 250 mL of water. Make up to 500 mL with water.
4. EDTA buffer: 0.5 M EDTA, pH 8.1. Weigh 146.12 g EDTA and transfer to a 1 L graduated cylinder containing approximately 500 mL of water. Adjust pH to 8.1 with NaOH (*see* **Note 13**). Make up to 1 L with water.
5. Protease inhibitor cocktail (PIC). Store at −20 °C.
6. Phenylmethylsulfonyl fluoride (PMSF). 87 mg/mL solution in 96 % EtOH (*see* **Note 14**). Store at −20 °C.
7. ChIP SDS lysis buffer: 1 % SDS, 10 mM EDTA, and 50 mM Tris, pH 8.1 (*see* **Note 15**).
8. Appropriate antibodies and controls (*see* **Note 16**).
9. ChIP buffer (5×) (we recommend One-Day ChIP kit from Diagenode, Liege, Belgium, based on our experience with this assay).
10. Deionized water.
11. Protease Inhibitor mix (200×) (Diagenode, Liege, Belgium).
12. Antibody binding beads (Diagenode, Liege, Belgium).
13. DNA purifying slurry (Diagenode, Liege, Belgium).
14. Proteinase K (Diagenode, Liege, Belgium).
15. Microcentrifuge tube mini floating rack.

2.6 RT-PCR Setup and Agarose Gel Electrophoresis

1. GoTaq® Green Master Mix (400 μM dATP, dGTP, dCTP, dTTP, 3 mM $MgCl_2$, GoTAQ, pH 8.5) (Promega, Madison, WI, USA: we recommend this vendor for the GoTaq mix due to superior results compared with other brands; it also includes a loading dye for convenience and ease of use).
2. 50× Tris–acetate–EDTA (TAE buffer): 2 M Tris, 50 mM EDTA, pH 8.0. Weigh 242 g Tris base and transfer to a 1 L glass bottle containing approximately 500 mL of water. Add 57.1 mL glacial acetic acid and 100 mL of 0.5 M EDTA (pH 8.0) buffer. Place a magnetic stir bar (flea) into the bottle and place on a magnetic stirrer with medium speed for

approximately 30 min until all of the Tris base has dissolved. Transfer into a graduated cylinder and make up to 1 L with water (*see* **Note 17**).

3. 6× loading buffer: Add 3.75 mL glycerol solution to 6.25 mL water and 0.25 % weight/volume (w/v) of bromophenol blue (*see* **Note 18**). Store at 4 °C. Use 2 μL of loading buffer per 10 μL of sample (*see* **Note 19**).
4. DNA ladder.
5. Electrophoresis set.

3 Methods

Allow cell lines to reach 60–70 % confluence (*see* **Note 3**) before adding drugs (as outlined in Subheading 2.1 above). Carry out all incubations at room temperature and centrifugation steps at 4 °C unless stated otherwise. All steps below should be performed in a laminar flow hood, except for the initial addition of the TRI reagent® to cells. Care must be taken when working with RNA. Designate a special area for RNA work only. Treat all surfaces with an RNase-inactivating agent; use RNase-free plastic and DEPC-treated water.

3.1 RNA Isolation

1. Decant medium from tissue culture flask and add 1 mL TRI reagent®. Incubate with agitation on a shaker for 5 min.
2. Scrape cells from the bottom of the flask using a cell scraper. Transfer TRI reagent® to a 1.5 mL microfuge tube.
3. Add 100 μL BCP to each sample, invert for 15 s, and incubate for 10 min. Centrifuge samples at 13,500 × *g* for 15 min.
4. Transfer colorless upper aqueous phase (containing RNA) to a clean 1.5 mL microfuge tube and discard the remaining phases (*see* **Note 20**).
5. Add 500 μL isopropanol (2-propanol) to each sample. Mix by inversion and incubate for 10 min. Centrifuge at 13,500 × *g* for 8 min.
6. Decant supernatant and add 1 mL EtOH wash buffer to the RNA pellet. Incubate for 5 min.
7. Centrifuge at 13,500 × *g* for 5 min and decant the EtOH wash buffer. Air-dry pellet for 5 min and resuspend in 50 μL water (*see* **Note 21**).
8. Quantify the RNA (*see* **Note 22**). Proceed to Subheading 3.2 or store at −80 °C until required.

3.2 RQ1 DNase Treatment

1. Transfer 1 μg RNA to a microfuge tube. Add 5 μL 10× RQ1 buffer and 1 μL RQ1 DNase to the RNA sample and adjust volume to 50 μL with water (*see* **Note 23**). Mix samples with a pipette tip and incubate at 37 °C for 1.5 h.

2. Add 50 μL phenol:chloroform:isoamyl alcohol solution to each sample. Vortex briefly and centrifuge at 13,500 × *g* for 5 min.
3. Transfer the upper layer containing the RNA to a clean microfuge tube. Add 1 μL polyacryl carrier, 50 μL 5 M ammonium acetate, and 300 μL 96 % EtOH. Vortex and incubate for 15 min.
4. Centrifuge at 13,500 × *g* for 20 min. Discard supernatant and wash pellet with 1 mL EtOH wash buffer (70 % solution). Centrifuge at 13,500 × *g* for 5 min.
5. Discard the EtOH wash buffer, and resuspend pellet in 10 μL water. Proceed to Subheading 3.3 or store at −80 °C until required.

3.3 cDNA Preparation

1. Add 1 μL 50 μM Oligo(dT)$_{20}$, 1 μL 10 mM dNTP stock, and 1 μL water to the 10 μL RNA sample from Subheading 3.2 (final reaction volume 13 μL). Mix sample with a pipette.
2. Incubate at 65 °C for 5 min and transfer to ice for 1 min.
3. Add 4 μL 5× FS buffer, 1 μL 0.1 M DTT, 1 μL ribonuclease inhibitor (40 U/μL), and 1 μL reverse transcriptase (200 U/μL) to each sample. Mix gently with a pipette and incubate at 50 °C for 1 h. Stop the reaction by incubating the samples at 70 °C for 15 min (*see* **Note 24**). Proceed to Subheading 3.6 or store samples at −20 °C until required.

3.4 Chromatin Immunoprecipitation Assay (X-ChIP)

When cells have been treated with the HDi (e.g., 250 ng/mL TSA) and vehicle control (untreated) for the appropriate period of time, the X-ChIP assay is carried out over 2 days. Volumes indicated here are for 20 immunoprecipitates (*see* **Note 25**), and are based on the One-Day ChIP Kit from Diagenode (Cat# kch-oneDIP-060) which is our preferred X-ChIP kit. Scale down accordingly for fewer samples. On day 1, the DNA is cross-linked, sheared, and incubated with appropriate antibody overnight while on day 2, the immunoprecipitation is completed and DNA purification steps are performed. The result is a sample of DNA ready for PCR. All centrifugation steps should be performed at 4 °C unless stated otherwise.

Day 1:

1. Add formaldehyde directly to the cell culture medium within the tissue culture flask to a final concentration of 1 % (270 μL of 37 % formaldehyde solution in 10 mL of culture medium). Mix thoroughly and incubate for 5 min at 37 °C. Decant medium and wash cells twice with ice-cold PBS.
2. Add 1 mL ice-cold PBS into the flask and scrape cells from the bottom of the flask using a cell scraper and transfer into a clean 1.5 mL microfuge tube. Centrifuge samples at 14,000 × *g* for 4 min.

3. Remove supernatant and resuspend pellet in 200 μL freshly prepared ChIP SDS lysis buffer (*see* **Note 26**). Incubate on ice for 30 min.
4. Sonicate samples while keeping on ice. Sonicate for a period of 10 s, followed by a rest of 30 s (*see* **Note 27**). Repeat three times.
5. Centrifuge for 10 min at 13,000 × *g*. Transfer 200 μL supernatant (sheared DNA) into a clean microfuge tube. Remove 10 μL supernatant for analysis (*see* **Note 28**).
6. Thaw antibodies and protease inhibitor mix on ice.
7. Prepare 1× ChIP buffer (*see* **Note 29**). Dilute the remaining sheared DNA tenfold (190 μL supernatant and 1.9 mL of buffer) into a 15 mL tube. This is the sheared chromatin for use in IP experiments.
8. Dilute sheared DNA to a maximum of 28 % sheared DNA relative to 1× ChIP + protease inhibitor buffer volume (e.g., 1.32 mL sheared DNA: 4.84 mL buffer) (*see* **Note 30**).
9. Transfer a 280 μL aliquot of the diluted sheared DNA into appropriately labelled 1.5 mL microfuge tubes (on ice) (*see* **Note 31**). Transfer 6 μL into a separate microfuge tube; this will serve as the input DNA (*see* **Note 32**) in later steps. Freeze excess at −80 °C.
10. Add antibody to appropriate tubes (refer to Table 1, *see* **Note 33**) leaving one set of tubes free of antibody to serve as the "no antibody" control (*see* **Note 34**) and vortex briefly. Incubate overnight at 4 °C on a rotary shaker.

Table 1
Antibodies used for X-ChIP assay

Antibody	Abbreviation	Supplier	Catalogue #
Pan acetylated histone H3	AcH3	Millipore	06-599
Pan acetylated histone H4	AcH4	Millipore	06-598
Histone H3 acetylated at lysine 9 and 14	H3K9/K14Ac	Diagenode	pAb-ACHBHS-044
Histone H3 acetylated at lysine 9	H3K9Ac	Diagenode	pAb-ACHAHS-044
Histone H3 acetylated at lysine 9 and phosphorylated at serine 10	H3K9Ac/S10p	Sigma-Aldrich	H0788
Histone H3 dimethylation marker at lysine 9	H3K9Me2	Sigma-Aldrich	D5567
Histone H3 dimethylation marker at lysine 4	H3K4Me2	Sigma-Aldrich	D5692
Histone H3 methylation marker at lysine 4	H3K4Me	Sigma-Aldrich	M4819

Day 2:

1. Resuspend antibody-binding beads and transfer 840 μL into 10.5 mL 1× ChIP buffer. Mix by inversion and centrifuge for 3 min at 500 × *g*. Remove supernatant and resuspend beads in 10.5 mL 1× ChIP buffer. This is enough for 20 individual IP reactions.
2. Transfer 500 μL of the washed beads into 1.5 mL microfuge tubes (*see* **Note 35**). Centrifuge for 2 min at 500 × *g*. Check by eye that equal pellets of beads are present in each tube. Remove supernatant (*see* **Note 36**). Store tubes on ice until required.
3. Centrifuge microfuge tubes containing antibody (from overnight incubation, day 1—**step 10**) for 10 min at 12,000 × *g*.
4. Transfer 250 μL each supernatant (*see* **Note 37**) to the pelleted beads (derived from **step 2** above). Incubate for 30 min at 4 °C on a rotary shaker.
5. Thaw input DNA aliquot (from day 1, **step 9**) and add 30 μL 96 % EtOH. Incubate on ice for 10 min and centrifuge at 10,000 × *g* for 10 min. Discard supernatant, and allow pellet to air-dry on ice. Resuspend pellet in 100 μL water and incubate at room temperature until ready to perform **step 10** below.
6. Add 1 mL ice-cold 1× ChIP buffer to all tubes from **step 4**. Invert twice. Centrifuge at 500 × *g* for 2 min.
7. Remove the supernatant and resuspend beads in 1 mL ice-cold 1× ChIP buffer. Transfer to a fresh 15 mL tube containing 12 mL cold ChIP buffer (*see* **Note 38**). Incubate at 4 °C for 5 min and centrifuge for 3 min at 500 × *g* at 4 °C.
8. Remove 12 mL buffer from each tube and resuspend beads in the 1 mL remaining buffer. Transfer to fresh 1.5 mL microtube (on ice). Centrifuge for 2 min at 500 × *g*.
9. Remove supernatant. Allow pellets to air-dry.
10. Resuspend the DNA-purifying slurry and add 100 μL to each bead pellet. Also add 100 μL DNA-purifying slurry directly to the input DNA sample from **step 5**. Incubate for 1 min. Invert tubes and incubate for 10 min in 100 °C water (*see* **Note 39**).
11. Cool samples and add 1 μL Proteinase K to each and vortex. Incubate for 30 min at 1,000 revolutions per minute (rpm) in a thermomixer at 55 °C.
12. Incubate in 100 °C water for 10 min (*see* **Note 39**). Centrifuge at 14,000 × *g*. The resulting supernatant is ready for subsequent PCR reactions (*see* **Note 40**).

3.5 Primer Design

Primer sequences for RT-PCR and ChIP are outlined in Tables 2 and 3, respectively. Primers used to study the promoter regions of *CXCL8*, and *CXCR1* and *CXCR2*, by ChIP were designed from the known promoters and/or 5′ UTRs contained within their

Table 2
Primers and annealing temperatures for RT-PCR

Primer (product size)	Sequence	Annealing temperature (°C)
CXCL1 (324 bp)	F: 5′-ATGGCCCGCGCTGCTCTCTC-3′ R: 5′-TCAGTTGGATTTGTCACTGTTC-3′	55
CXCL2 (324 bp)	F: 5′-ATGGCCCGCGCCACGCTCTC-3′ R: 5′-TCAGTTGGATTTGCCATTTTTCAGC-3′	55
CXCL3 (324 bp)	F: 5′-ATGGCCCACGCCACGCTCTCCG-3′ R: 5′-TCAGTTGGTGCTCCCCTTGTTC-3′	60
CXCL8 (297 bp)	F: 5′-ATGACTTCCAAGCTGGCCGTG-3′ R: 5′-TGAATTCTCAAGCCCTCTTCA-3′	55
CXCR1 (300 bp)	F: 5′-CCTTCTTCCTTTTCCGCCAG-3′ R: 5′-AAGTGTAGGAGGTAACACGATG-3′	60
CXCR2 (512 bp)	F: 5′-CTGCTCTGCTGGCTGCCCTA-3′ R: 5′-GAGAGTAGTGGAATTGTGCCC-3′	60
Beta-actin (510 bp)	F: 5′-TGTTTGAGACCTTCAACACCC-3′ R: 5′-AGCACTGTGTTGGCGTACAG-3′	55

Table 3
Primers and annealing temperatures for ChIP

Primer (product size)	Sequence	Annealing temperature (°C)
CXCL8 (171 bp)	F: 5′-AAGAAAACTTTCGTCATACTCCG-3′ R: 5′-TGGCTTTTTATATCATCACCCTAC-3′	55
CXCR1 (120 bp)	F: 5′-TCAACCCCAGCTTCACACCT-3′ R: 5′-CCCCACCATGTCTACATTT-3′	56
CXCR2 (215 bp)	F: 5′-GGAAACTTCCCACCAGTGAA-3′ R: 5′-CCACTGGATCCCTTGCAGAC-3′	60

nucleotide sequences (*see* **Note 41**). Tables include product sizes and annealing temperatures. Cycling conditions are as follows: Template DNA is initially denatured at 94 °C for 5 min, followed by 35 amplification cycles in a thermal cycler. Each cycle consists of template denaturation (1 min at 94 °C), primer annealing (1 min at target annealing temperature), and extension (1 min at 72 °C). This is followed by an elongation step (72 °C for 10 min) to complete the amplification cycle (*see* **Note 42**).

3.6 Reverse Transcriptase Polymerase Chain Reaction

cDNA (Subheading 3.3) or DNA from the ChIP assay (Subheading 3.4) is used as a template for the PCR.

1. Prepare the following PCR master-mix for each sample to be amplified: 10 μL 2×GoTaq® Green Master Mix, 2 μL 5 μM forward primer, 2 μL 5 μM reverse primer, and 6 μL water (final reaction volume 20 μL).
2. Use 1 μL cDNA template per reaction (2.5 μL for ChIP and adjust water accordingly). Include a negative control for each RT-PCR reaction (*see* **Note 43**).

3.7 Agarose Gel Electrophoresis

All PCR products are resolved by electrophoresis on a 2 % agarose gel (*see* **Note 44**) containing ethidium bromide (*see* **Note 45**) and visualized with a UV gel documentation system (*see* **Note 46**).

1. Weigh 2 g agarose and dissolve in 100 mL 1× TAE buffer by boiling in a microwave for 2–3 min on a medium heat setting. Check periodically and swirl repeatedly.
2. Allow the solution to cool down to 55–60 °C (hand hot) before adding ethidium bromide to achieve a final concentration of 1 μg/mL (10 μL in 100 mL).
3. Pour the 2 % agarose solution into a gel tray with well-forming combs to a depth of 3–5 mm and allow to solidify (approximately 20–30 min). Once solidified, pour sufficient 1×TAE buffer to the maximum level as indicated on gel rig. Remove combs.
4. Load a 10 μL aliquot of each PCR product (including negative) to individual wells (*see* **Note 19**). Mix 1 μL of the DNA ladder with 5 μL loading buffer and load onto the gel. Keep voltage constant at 120 V and allow electrophoresis to run for approximately 30–40 min until dye front is approximately ¾ of the way down the gel. Visualize and photograph under UV light and perform densitometry analysis (*see* **Note 47**).
5. Figure 2a, b is provided for illustrative purposes. Figure 2a: Expression of CXC (ELR$^+$) genes after treatment with (a) TSA and (b) DAC in A549 cells (a non-small-cell lung cancer cell line). Figure 2b: ChIP assay for histone acetylation in A549 cells following TSA treatment. Figures are provided through the Creative Commons Attribution License and are from Baird et al. [20].

4 Notes

1. There are a variety of drugs and chemicals available to study epigenetic mechanisms. Individual drugs may require different solvents and may be soluble only at specific concentrations, which will affect the concentration of the drug stock.

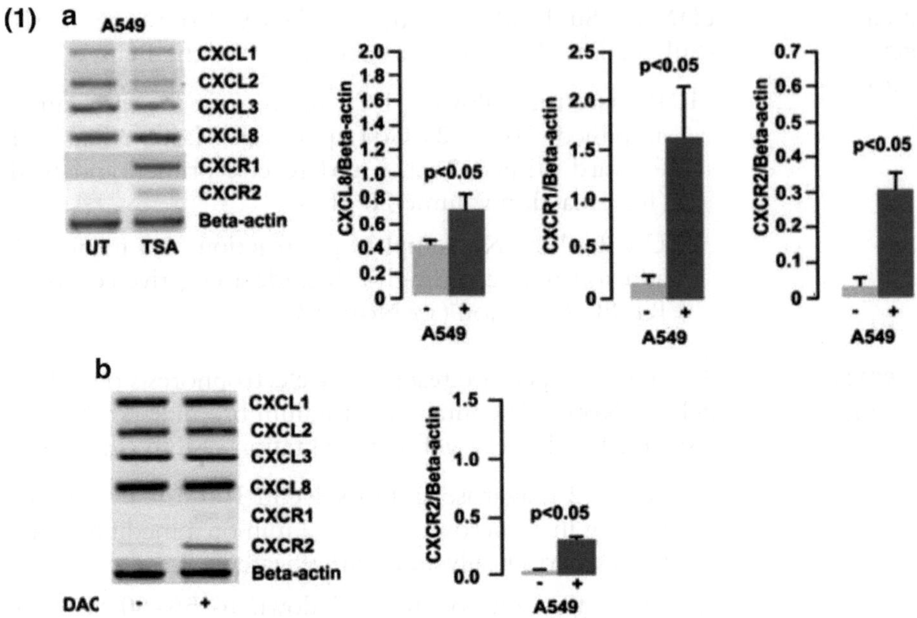
(1)
a
A549
CXCL1
CXCL2
CXCL3
CXCL8
CXCR1
CXCR2
Beta-actin
UT TSA
CXCL8/Beta-actin
CXCR1/Beta-actin
CXCR2/Beta-actin
p<0.05
A549
b
DAC
1.5
1.0
0.5
0.0

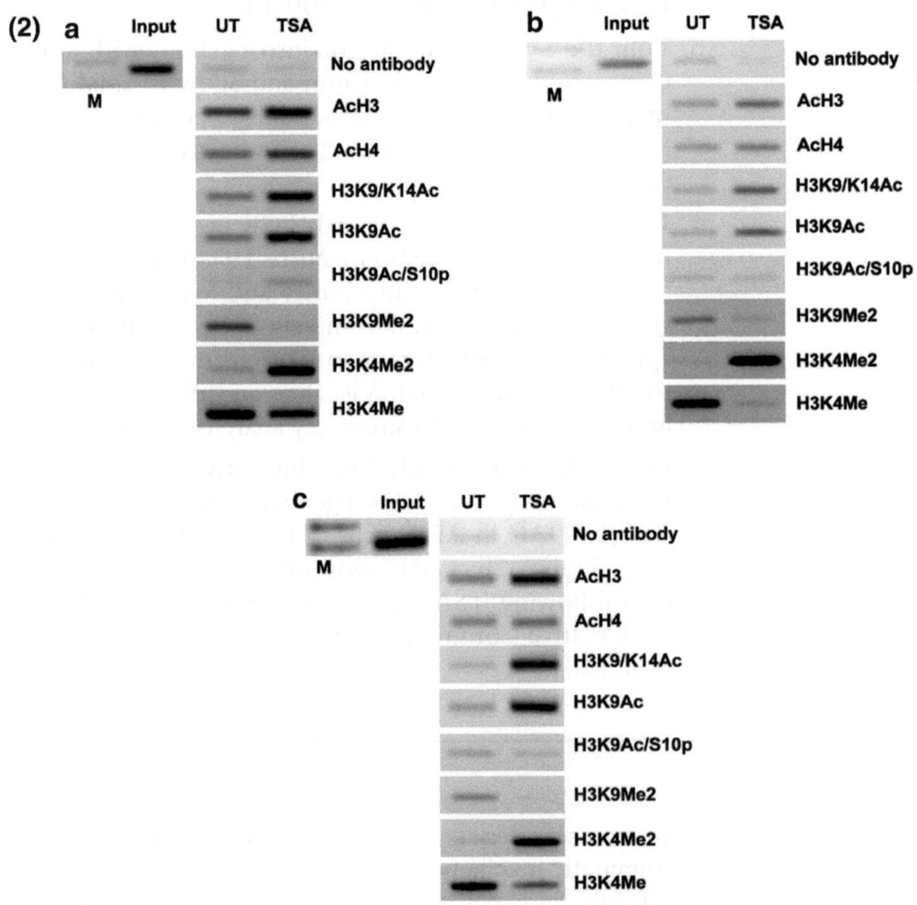
(2)
a
b
c
Input
UT
TSA
M
No antibody
AcH3
AcH4
H3K9/K14Ac
H3K9Ac
H3K9Ac/S10p
H3K9Me2
H3K4Me2
H3K4Me

2. If a drug is dissolved in a particular solvent, then the vehicle control (often referred to as untreated) must be the same solvent. For instance treating a flask of cells with TSA (dissolved in DMSO) will require 10 μL TSA stock in 10 mL culture medium. The appropriate untreated control will be 10 μL DMSO in 10 mL of culture medium.
3. In our laboratory, non-small-cell lung cancer cell lines are cultured in cell-specific medium containing any necessary supplements and antibiotics (such as penicillin–streptomycin) and 10 % fetal bovine serum (FBS). To determine the effect of any chemicals or drugs, cells are first seeded in complete medium containing 10 % FBS. When the appropriate cell confluence is reached, the cells are washed with PBS and complete medium with 0.5 % FBS is added to the flask for a period of 24 h prior to the addition of chemicals or drugs. We refer to this as serum starvation. All treatments are undertaken in complete medium with 0.5 % FBS.
4. It is of the utmost importance to frequently screen cell lines for mycoplasma contamination (every 3–6 months) as this can affect cell characteristics. In our laboratory we use a PCR-based method adapted from Young et al. [21].
5. In some cell lines, the effect of TSA can be cell density dependent, so it is important to treat cells during the logarithmic phase of growth [22]. To minimize variability, we initially plate cells at the same density for each experiment.
6. The drug and medium must be replaced every 24 h. This is to allow for the incorporation of any DNA methylation changes through the cell proliferation cycle, and to allow for DNA methyltransferase inhibition (DNMTi) degradation.
7. When taking an aliquot, make sure to go below the upper layer into the organic phase.
8. Having a small amount of the appropriate solvent at the bottom of a bottle or a graduated cylinder helps to more quickly dissolve most chemicals. Make sure that the entire chemical has dissolved before making the solution up to its final volume.
9. Polyacryl carrier intercalates with RNA, allowing for easier identification of the RNA pellet. Adding this carrier can affect the optical density of RNA. To normalize for this effect,

Fig. 2 (1) The effect of (a) TSA and (b) DAC treatment on the expression of CXCL1-3, CXCL8, CXCR1, and CXCR2 on A549 cells. Densitometry analysis of chemokine ligand and receptor expression in treated versus untreated samples was normalized to beta-actin. Semiquantitative data are graphed as means ± standard error of mean ($n = 3$). Statistical comparisons were performed using a one-tailed student's t-test [20]. The figure is provided under the Creative Commons Attribution License. **(2)** Histone acetylation in ELR^{+} CXC chemokine ligand and receptors in A549 cells following TSA treatment. Histone acetylation occurred directly at the promoter regions of (a) CXCL8, (b) CXCR1 and (c) CXCR2 in the A549 cells after treatment with TSA [20]. The figure is provided under the Creative Commons Attribution License

process a blank sample containing only the reagent (in this case, water) and the polyacryl carrier, when quantifying RNA samples.

10. We use Oligo(dT)$_{20}$ (homogenous mixture of 20-mer thymidines) for cDNA synthesis as it has a great specificity for mRNA and allows many different targets to be studied from the same cDNA pool.
11. dTTP, dATP, dGTP, and dCTP are usually supplied as individual 100 mM stocks. To make a 10 mM dNTP stock add 10 μL of each to 60 μL of water (final volume 100 μL).
12. SDS can precipitate out of solution at colder temperatures. Therefore the buffer must be pre-warmed before use.
13. NaOH is essential for the solubility of EDTA. We generally make a dilute (0.5 M—for fine-tuning the pH) and concentrated (6 M—for making large changes) NaOH solution to adjust the pH. We also keep the EDTA solution stirring on a magnetic stirrer to ensure uniform NaOH distribution when measuring and adjusting pH.
14. PMSF is a supersaturated solution that requires vigorous vortexing.
15. A simple way to prepare lysis buffer is as follows: Measure 5 mL of 10 % SDS buffer, 5 mL of 100 mM Tris buffer, and 5 mL of 0.5 M EDTA buffer into a graduated cylinder. Make up to 50 mL with water.
16. Decide which targets are going to be investigated such as acetylated histone H3 and procure ChIP appropriate antibodies. The negative controls can consist of an IgG antibody raised in the same species as the ChIP antibodies or a "no antibody" control to detect for nonspecific binding.
17. It is more convenient to make a 50× stock of TAE buffer and dilute to 1× as required (e.g., 20 mL of 50× stock in 1 L of water). The 0.5 M EDTA buffer is made as mentioned previously (Subheading 2.5, **item 4**) and divided into two separate aliquots, one adjusted to pH 8.1 for the ChIP assay (Subheading 2.5) and one that is adjusted to pH 8.0 for the TAE buffer.
18. Glycerol is a viscous liquid and difficult to pipette. Glycerol can be pre-warmed in a microwave for approximately 30 s, or cut a pipette tip at an angle to increase the bore size or measure using a plastic syringe with no needle.
19. We find that it is easier to mix PCR product with loading buffer (where appropriate) on a piece of Parafilm®. Cut a piece of Parafilm® to required length and sanitize with 70 % ethanol. Pipette 2 μL aliquot loading buffer onto the Parafilm® and add 10 μL of PCR product, mixing a number of times with pipette

tip before loading onto agarose gel. There are a number of premade Taq mixes currently available which include a loading dye such as the GoTaq® Green Master Mix we use.

20. TRI reagent® can be used for the isolation of RNA, DNA, and protein from a single sample. After the initial centrifugation step (subsequent to adding BCP), the sample separates into three phases. The upper colorless aqueous phase contains RNA, while the interphase (middle) contains DNA and the organic lower phase contains protein. It is advised to leave some of the upper phase behind to avoid RNA contamination.
21. The volume of molecular grade water will vary depending on the RNA pellet size. For smaller pellets, use about 20 μL water.
22. The 260:280 absorbance ratio should be approximately 2. If this ratio is lower than this, protein or phenol contamination or both should be considered.
23. A master mix can be prepared (containing buffer, RQ1 DNase) for all of the reactions. Always make enough master mix for one or two additional samples to control for pipette error. Master mixes can also be prepared for the cDNA and RT-PCR protocols.
24. The temperature and incubation times will vary depending on the type of reverse transcriptase that is used. The temperatures and times given here refer to Superscript™ III reverse transcriptase (Invitrogen Corp., Carlsbad, CA, USA).
25. Twenty immunoprecipitates include samples from two treatments (e.g., untreated, TSA), incubated with eight antibodies each, and a no antibody control, and then a positive control and an extra sample to control for pipette error.
26. Warm ChIP SDS lysis buffer (e.g., Millipore Cat# 20-163) to room temperature before use and add the following: 10 μL PIC and 2 μL PMSF to 1 mL buffer.
27. Sonicate samples (to shear DNA) on a medium setting (refer to manufacturers' instructions) by placing needle into the tube just above the bottom. Be careful to avoid contact of the needle with tube surface. Allow sample adequate time to cool between sonications.
28. Run sample on an agarose gel to determine shearing efficiency. If the DNA is inadequately sheared, repeat the sonication steps. The duration and the number of sonication cycles must be optimized for each cell line before continuing with the rest of the ChIP protocol.
29. Dilute ChIP buffer (5×) to a 1× stock in deionized water (add 100 mL ChIP buffer to 400 mL water). Store at 4 °C.

30. For example for one IP—66 μL sheared DNA: 242 μL buffer.
31. For example, if you have an untreated and a treated sample and are testing two antibodies you will need four tubes, plus tubes for controls.
32. Input DNA serves as a positive control and will be used at a later stage in the protocol.
33. You will need to use approximately 1–2 μg of antibody. We have successfully used 2 μg of antibody in these experiments.
34. A "no antibody" control is used to test for nonspecific binding. These tubes will be treated in the same way as the others, but will not contain any antibody.
35. The beads must be kept in suspension. Mix beads in between each aliquot. Use a tip that has been cut at an angle.
36. Aspirate supernatant slowly. Do not disturb the pellet. If pellet is disturbed, centrifuge the samples again. It is important to check if any buffer remains in the cap and if present remove.
37. Do not disturb the pellet, as this contains the unspecific aggregates. Do not add all of the supernatant antibody–chromatin mix to the pellet of beads. Do not take more than 250 μL from the tubes and discard the excess.
38. Pipette beads against the side of the tube at the top of the liquid as this helps to wash beads further as they sediment through the liquid.
39. Many different cap locks are available to prevent the cap from popping open due to the heat.
40. Additional information on the Diagenode One-day ChIP Kit protocol can be found at http://www.diagenode.com/media/catalog/file/OneDay_ChIP-kit_manual.pdf.
41. A number of different software packages are available including free online versions to help with primer design.
42. It is helpful to use a positive control sample to optimize PCR cycling conditions. Ensure that one clean band is present at the expected product size.
43. It is imperative to include a negative sample with each PCR performed. This should consist of all the components of the PCR reaction with water in place of template DNA. No band should appear in the negative control well.
44. The percentage of agarose gel used depends on the PCR product size. Use a 2 % gel (2 g of agarose in 100 mL 1× TAE) for products below 1 kb.
45. Ethidium bromide is a carcinogenic agent and requires careful handling and proper disposal. Compliance with all local rules regarding its use is mandatory.

46. PCR products were visualized and photographed under UV light using a Biospectrum Imaging System (Ultra Violet Products, Cambridge, UK) in our laboratory.
47. We load sample (e.g., CXCL1) (10 μL) and control (beta-actin) (2 μL) side by side on the one gel. Densitometry analysis was performed using the TINA 2.09c densitometry program (Raytest, Germany). The target gene expression is normalized to the control gene and expressed as ratio of target expression:control gene expression.

References

1. Waddington CH (2012) The Epigenotype. Reprinted in Int J Epidemiol 41:10–13
2. Baylin SB, Jones PA (2011) A decade of exploring the cancer epigenome - biological and translational implications. Nat Rev Cancer 11:726–734
3. Collas P (2010) The current state of chromatin immunoprecipitation. Mol Biotechnol 45:87–100
4. Carey MF, Peterson CL, Smale ST (2009) Chromatin immunoprecipitation (ChIP). Cold Spring Harb Protoc. doi:10.1101/pdb.prost5279
5. Gilmour DS, Lis JT (1984) Detecting protein-DNA interactions in vivo: distribution of RNA polymerase on specific bacterial genes. Proc Natl Acad Sci USA 81:4275–4279
6. Gilmour DS, Lis JT (1985) In vivo interactions of RNA polymerase II with genes of Drosophila melanogaster. Mol Cell Biol 5:2009–2018
7. Gilmour DS, Lis JT (1986) RNA polymerase II interacts with the promoter region of the noninduced hsp70 gene in Drosophila melanogaster cells. Mol Cell Biol 6:3984–3989
8. Jackson V (1978) Studies on histone organization in the nucleosome using formaldehyde as a reversible cross-linking agent. Cell 15:945–954
9. Solomon MJ, Larsen PL, Varshavsky A (1988) Mapping protein-DNA interactions in vivo with formaldehyde: evidence that histone H4 is retained on a highly transcribed gene. Cell 53:937–947
10. O'Neill LP, Turner BM (2003) Immunoprecipitation of native chromatin: NChIP. Methods 31:76–82
11. Turner, B. (2001) ChIP with Native Chromatin: Advantages and Problems Relative to Methods Using Cross-Linked Material. In: Mapping Protein/DNA Interactions by Cross-Linking., Institut national de la sante et de la recherche medicale (INSERM): Paris.
12. Hebbes TR, Thorne AW, Crane-Robinson C (1988) A direct link between core histone acetylation and transcriptionally active chromatin. EMBO J 7:1395–1402
13. Boyd KE, Farnham PJ (1997) Myc versus USF: discrimination at the cad gene is determined by core promoter elements. Mol Cell Biol 17:2529–2537
14. Grandori C, Mac J, Siebelt F, Ayer DE, Eisenman RN (1996) Myc-Max heterodimers activate a DEAD box gene and interact with multiple E box-related sites in vivo. EMBO J 15:4344–4357
15. O'Neill LP, VerMilyea MD, Turner BM (2006) Epigenetic characterization of the early embryo with a chromatin immunoprecipitation protocol applicable to small cell populations. Nat Genet 38:835–841
16. Dahl JA, Reiner AH, Collas P (2009) Fast genomic muChIP-chip from 1,000 cells. Genome Biol 10:R13
17. Hecht A, Strahl-Bolsinger S, Grunstein M (1996) Spreading of transcriptional repressor SIR3 from telomeric heterochromatin. Nature 383:92–96
18. Mukhopadhyay A, Deplancke B, Walhout AJ, Tissenbaum HA (2008) Chromatin immunoprecipitation (ChIP) coupled to detection by quantitative real-time PCR to study transcription factor binding to DNA in Caenorhabditis elegans. Nat Protoc 3:698–709
19. Wong E, Wei CL (2009) ChIP'ing the mammalian genome: technical advances and insights into functional elements. Genome Med 1:89
20. Baird AM, Gray SG, O'Byrne KJ (2011) Epigenetics underpinning the regulation of the CXC (ELR^+) chemokines in non-small cell lung cancer. PLoS One 6:e14593
21. Young L, Sung J, Stacey GA, Masters JR (2010) Detection of Mycoplasma in cell cultures. Nat Protoc 5:929–934
22. Gray SG, Ekstrom TJ (1998) Effects of cell density and trichostatin A on the expression of HDAC1 and p57Kip2 in Hep 3B cells. Biochem Biophys Res Commun 245:423–427

Chapter 13

Using Fluorescent Chemokine Uptake to Detect Chemokine Receptors by Fluorescent Activated Cell Sorting

Laura B. Ford, Chris A.H. Hansell, and Robert J.B. Nibbs

Abstract

Fluorescent activated cell sorting (FACS) is a powerful technique that allows rapid quantitative cell-by-cell analysis of the expression of multiple distinct proteins and the isolation of specific cell types from complex cellular mixtures. FACS-based detection of chemokine receptors is typically achieved using chemokine receptor-specific antibodies. However, this approach has some limitations, and, in our experience, many anti-chemokine receptor antibodies have poor signal-to-noise ratios and thus low sensitivity and reliability in FACS. Moreover, antibodies against some chemokine receptors are not commercially available, particularly in nonhuman species, and generating good antibodies against chemokine receptors is challenging and expensive. Here, we describe a simple alternative method of detecting chemokine receptors by FACS that relies on cellular internalization of fluorescently labelled chemokines. This approach detects chemokine receptors with high sensitivity, specificity, and reliability, and consistently outperforms antibody-based detection methods when applied to the analysis of mouse cells in our experience. It can also be used to study receptor specificity on primary cells and, since chemokines from one species often bind to receptors from other species, a single fluorescently labelled chemokine should allow reliable chemokine receptor detection and analysis across a range of species.

Key words Chemokine receptor, Fluorescent chemokine, Leukocyte, Antibody, Flow cytometry, FACS

1 Introduction

There are at least 18 conventional G-protein-coupled chemokine receptors in mammals, each able to bind one, or more, of the ~45 chemokines in these species [1]. They provide directional cues by which motile cells can navigate into, within, and between tissues. There are also four "atypical" chemokine receptors that are unable to direct cell migration, but which are instead thought to act as specialized chemokine scavengers that regulate chemokine biodistribution and abundance [2–4]. Few conventional, and no atypical, chemokine receptors bind just one chemokine ligand, and some chemokine receptors show highly promiscuous ligand binding

Astrid E. Cardona and Eroboghene E. Ubogu (eds.), *Chemokines: Methods and Protocols*, Methods in Molecular Biology, vol. 1013, DOI 10.1007/978-1-62703-426-5_13,

capabilities. Similarly, individual chemokines are rarely only recognized by a single chemokine receptor, and some have the capacity to bind four or more distinct receptors [1–4].

A full understanding of each chemokine receptor requires detailed insight into where and when it is expressed in the species under study. Transcript detection, by quantitative reverse transcriptase polymerase chain reaction (Q-RT-PCR) for instance, has its uses, but methods that detect chemokine receptor protein are clearly preferable, particularly if they work at the level of individual cells. This is usually achieved using chemokine receptor antibodies in immunohistochemistry, immunofluorescence, or fluorescent activated cell sorting (FACS). However, there are limitations with these approaches. First, there is a limited commercial supply of chemokine receptor antibodies, particularly for use in nonhuman species. In our experience, effective chemokine receptor antibodies are not easy to generate, and those raised against chemokine receptor-derived linear peptides often fail to detect native protein in tissue sections or cell suspensions. Posttranslational and conformational variation in chemokine receptor structure may also restrict immunoreactivity against native protein.

Second, antibodies raised against a chemokine receptor from one species uncommonly detect chemokine receptors in other species. Thus, individual antibodies are usually required to detect the same chemokine receptor across species. Third, although isotype-matched antibodies are used to control for nonspecific binding of the test antibody, they do not control for cross-reaction of the antibody with other proteins. Indeed, by comparing cells or tissue sections from wild-type (WT) and chemokine receptor-deficient mice, we have found that the "specific" detection of chemokine receptor on WT cells (suggested by comparing the test antibody to isotype-matched controls) is often not lost from chemokine receptor-deficient cells [5] (Ford et al., unpublished observations). Cells or tissues from chemokine receptor-deficient mice are sometimes included as negative controls in published studies (and should, in our view, be used more regularly where possible), but clearly such controls are not possible in studies conducted in species other than mice.

We study atypical chemokine receptors, particularly D6. We generated anti-human D6 monoclonal antibodies, and used them to show D6 expression by lymphatic endothelial cells and trophoblasts in this species [6, 7]. We have had less success producing effective anti-mouse D6 antibodies and, in our hands, those that are commercially available are unable to detect D6 by FACS or immunohistochemistry [5] (unpublished data). To overcome this problem, we developed the FACS-based assay described herein to determine whether chemokine scavenging by D6 could be exploited to detect mouse leukocytes with D6 "activity" (i.e., D6-mediated chemokine uptake) [5, 8]. In this simple assay, primary cell suspensions are incubated ex vivo at 37 °C with fluorescently tagged D6-binding human chemokine, washed, immunostained with fluorescently labelled antibodies at 4 °C (to enable the identification of specific leukocyte subsets), washed again, and then analyzed by FACS. During the 37 °C incubation, cells internalize the fluorescent chemokine by chemokine receptor-dependent and -independent mechanisms: key controls allow the mechanism of chemokine uptake, and the specific receptor responsible, to be defined. Using this assay we found that, amongst leukocytes, D6 activity is restricted to innate-like B cells in mice (i.e., marginal zone and B1 B cells) [5]. We also noticed that conventional chemokine receptors mediate extensive chemokine uptake [5].

Indeed, the fluorescent human chemokines we have tested so far provide a level of sensitivity and specificity in the FACS-based detection of mouse CCR2, CCR4, CCR5, CCR7, CXCR2, and CXCR3 that even the best anti-mouse chemokine receptor antibodies we have studied have difficulty achieving [5, 8] (Ford et al., unpublished observations). The same fluorescent human chemokines sensitively detect chemokine receptors expressed by rat and human leukocytes (Ford et al., unpublished observations), and we believe they are likely to be effective in studies on other mammals, and possibly nonmammalian species. Moreover, competition experiments, in which unlabelled chemokines are included as competitors during the 37 °C incubation step, allow receptor specificity on primary cells to be studied by FACS [5] (Ford et al., unpublished observations).

Immunostaining and fluorescent chemokine uptake can identify chemokine receptor-bearing cells by FACS, but there are key mechanistic differences that should be considered when deciding on which method to use, and when interpreting the data generated. Immunostaining is typically performed at 4 °C, so it only detects surface chemokine receptor. The 37 °C incubation step was included in the fluorescent chemokine assay because incubations at 4 °C gave poor surface chemokine receptor detection (although tetramers of biotinylated chemokines and fluorescent streptavidin were more reliable [unpublished observations]) and because we were specifically interested in chemokine scavenging in our D6 work [5]. At 37 °C, chemokine uptake "activity" will be influenced by surface receptor levels but also by the specificity, affinity, internalization rate, and intracellular trafficking properties of the receptor. Receptors transiting the cell surface during the incubation period are theoretically capable of mediating fluorescent chemokine uptake.

Fluorescent chemokine uptake is a more complex, dynamic process than immunostaining at 4 °C, and differences in this process between cells might reflect variation in one or more of the parameters above. The biological significance of such differences is clear when considering chemokine scavenging by atypical chemokine receptors like D6, but what they mean for the cellular function of conventional chemokine receptors remains uncertain. Nonetheless, the fluorescent chemokine uptake assay provides a sensitive, specific, cross-species alternative to antibody-based methods of chemokine receptor detection by FACS, which we believe has the potential to expand our understanding of the biology of chemokines and their receptors.

In this method of chemokine receptor detection, primary cells are incubated *ex vivo* with fluorescently labelled chemokine. Cells expressing chemokine receptors are expected to internalize the chemokine and accumulate measurable intracellular fluorescence, which can be compared by FACS to equivalent cell preparations that have been incubated without fluorescent chemokine. However, some cells might internalize chemokine without using specific chemokine receptors (e.g., by pinocytosis) and chemokine receptors often exhibit overlapping ligand binding profiles. Thus, it is essential that appropriate controls are selected to determine which chemokine receptors, if any, are responsible for fluorescent chemokine internalization.

This can be achieved by comparing the fluorescent signal detected from WT cell populations to that detected by the same population derived from mice with a genetic deletion of the predicted relevant chemokine receptor (e.g., a chemokine receptor knockout (KO) mouse). Loss of fluorescence in the KO sample indicates that the chemokine receptor deleted was most likely responsible for the fluorescent chemokine uptake observed using WT cells. We have used this approach to determine whether uptake of AlexaFluor-647-labelled human CCL2 ($CCL2^{AF647}$) was mediated by CCR2 or D6 (Fig. 1), both of which are CCL2 receptors [5] (Ford et al., unpublished observations).

In addition, and when KO animals are unavailable, the inclusion of unlabelled chemokines into control WT samples can (1) identify the receptors responsible for uptake by these cells, and (2) assess the level of nonspecific chemokine uptake. In these competition experiments, the unlabelled chemokine, which we usually add at tenfold molar excess or more, prevents receptor-mediated uptake of the fluorescent chemokine. Homologous competition, when the same chemokine is used as the labelled and unlabelled chemokine, will block all receptor-mediated uptake, but will not prevent nonspecific pinocytic accumulation of fluorescent chemokine. Heterologous competition, when the unlabelled and labelled chemokine are different but bind with high affinity to the same receptor, allows uptake via specific receptors to be defined. In this case, careful selection of fluorescent chemokine and competitor is required so that a unique pattern of receptor specificity provides definitive identification of the activity of a specific receptor.

Unlabelled chemokines derived from the species under study are preferable, but chemokines from other species can also be used. For example, human $CCL2^{AF647}$ can be internalized by mouse CCR2 and D6, but only D6 can bind mouse CCL22 and several other chemokines (e.g., CCL11, CCL17) [3, 5]. Thus, measurable competition of $CCL2^{AF647}$ internalization by an excess of unlabelled

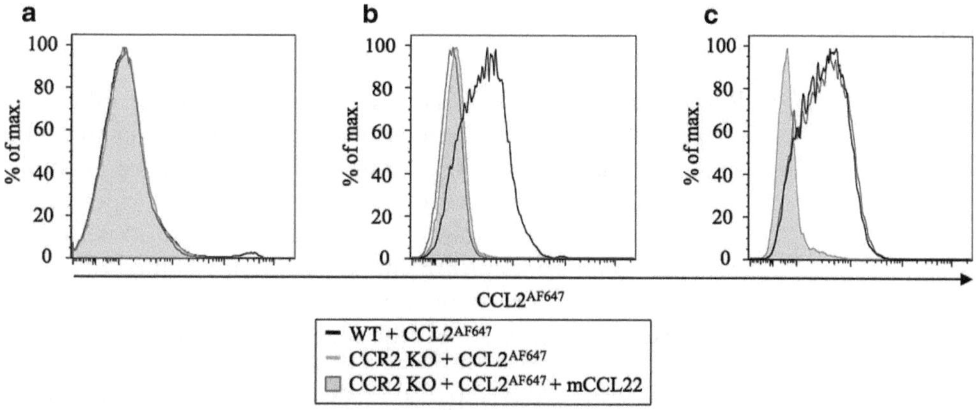

Fig. 1 Representative histogram overlays of FACS data generated after $CCL2^{AF647}$ uptake by specific splenocyte populations. $CCL2^{AF647}$ uptake assays were performed on WT (*black line*) and CCR2 KO (*grey line*) splenocytes. A control CCR2KO splenocyte sample (*filled grey plot*) contained $CCL2^{AF647}$ and a tenfold molar excess of mouse CCL22 (mCCL22), which binds D6 but not CCR2, so it only blocks D6-mediated uptake. After $CCL2^{AF647}$ uptake, splenocytes were stained with antibodies, and appropriately gated during analysis, to identify cell subsets in which the majority of the cells either did not internalize $CCL2^{AF647}$ using CCR2 or D6 (**a**) or internalized $CCL2^{AF647}$ using CCR2 (**b**) or D6 (**c**) only. Note in (**b**) there is a decrease in uptake in CCR2 KO compared to WT, but no further decrease with CCL22 competition, while in (**c**) there is no difference in $CCL2^{AF647}$ uptake between WT and CCR2 KO cells, but mouse CCL22 competition causes a reduction in $CCL2^{AF647}$ uptake by CCR2 KO cells [and WT cells (not shown)]. D6-mediated $CCL2^{AF647}$ uptake by cells in (**c**) was confirmed by observing its loss from equivalent populations of D6 KO splenocytes (not shown)

mouse CCL22 is indicative of D6 activity. Human CCL22 is also effective at distinguishing D6 from CCR2. Conversely, human $CCL2^{AF647}$ internalization by CCR2 can be competed by an excess of unlabelled mouse CCL7, which does not bind mouse D6 [5]. It is possible to identify similar combinations of labelled and unlabelled chemokine that identify, or exclude, specific chemokine receptors as being responsible for any observed fluorescent chemokine uptake [1] and we have been able to do this for all receptors we have studied (CCR2, CCR4, CCR5, CCR7, CXCR2, and CXCR3) [5, 8] (Ford et al., unpublished observations). Thus, a typical experiment might include samples containing (1) no fluorescent chemokine, (2) fluorescent chemokine alone, (3) fluorescent chemokine plus a tenfold molar excess of the same chemokine but unlabelled, and (4) fluorescent chemokine plus a tenfold molar excess of an unlabelled chemokine selected for its ability to only compete with the fluorescent chemokine for binding to a single receptor. Other samples containing fluorescent chemokine and a tenfold molar excess of other unlabelled chemokines might also be included to further confirm receptor specificity. Once primary cells expressing a specific receptor have been identified, a detailed profile of the specificity of that receptor can be generated and, by using a range of concentrations of unlabelled competitor, some insight gained into the relative activity of the competitor chemokines [5].

2 Materials

Recommended vendors are suggested based on the efficacy of the reagents in these assays based on our experience. Prepare all buffers and chemokine stocks in a sterile manner. Prepare all biological materials in a tissue culture hood using sterilized pipettes, pipette tips, and dissection equipment. Buffers and solutions can be prepared at room temperature and subsequently chilled or warmed as indicated in the protocol. For long-term storage all buffers and chemokine solutions should be maintained at 4 °C.

2.1 Buffers

1. Complete RPMI: RPMI 1640 medium supplemented with 10 % fetal calf serum (FCS), 10 U/ml penicillin/streptomycin, and 0.2 mM l-glutamine.
2. Collagenase D solution: Collagenase D (e.g., Roche, UK.) 0.1 mg/ml in HBSS without calcium or magnesium (e.g., Invitrogen, UK).
3. Binding buffer: Complete RPMI containing 20 mM HEPES (prepared as a 1:50 dilution of a 1 M stock solution).
4. FACS buffer: 1× PBS, containing 1 % FCS, 0.02 % sodium azide (NaN_3), and 5 mM EDTA (pH 7.4). This buffer can be made in advance and stored at 4 °C for several weeks.

2.2 Chemokines

1. AlexaFluor® 647 (AF647) labelled chemokines from Almac, UK. This vendor chemically synthesizes their chemokines with fluorophores conjugated to a single site (the C-terminal residue). Chemokines made in this fashion are advantageous as they are pure, homogeneous, highly active, and endotoxin-free. These chemokines are re-suspended in sterile PBS to a final concentration of 5 μM and stored at 4 °C in the dark. Alternatively, and following the manufacturer's recommendations, chemokines can be reconstituted in distilled water or dimethylsulfoxide (DMSO) in water (up to 30 % DMSO volume/volume). The manufacturer recommends that lyophilized products can be stored for at least 24 months. We find that reconstituted products can be stored at 4 °C for several weeks without affecting the quality of the product. The reconstituted product can also be stored at −20 °C for several months, but repeated freeze–thaw cycles should be avoided.
2. Unlabelled murine chemokines (e.g., Peprotech, UK, or R&D Systems, UK) are re-suspended in sterile, endotoxin-free H_2O at a final concentration of 10 μM and stored at −20 °C until use. Repeated freeze–thaw cycles should be avoided, but, in our experience, chemokine aliquots can be stored at −20 °C for at least 6 months without loss of activity. In our experience, these vendors are the best source for chemokines that are of high purity and biological activity.

2.3 Antibody Staining

1. Antibodies for detecting specific cell types, and appropriate isotype controls, can be purchased from any reputable supplier (e.g., BD Biosciences, UK, or eBiosciences, UK). These are directly coupled to fluorophores, biotinylated (detected with fluorescent streptavidin), or unconjugated (and detected using fluorescently labelled secondary antibodies that specifically recognize the primary antibody).
2. Fc block (e.g., BD Biosciences, UK) is used to reduce nonspecific antibody binding to cells.

2.4 Additional Materials

The following materials are also required:

1. Sterile dissection equipment (e.g., scalpel, tweezers).
2. Mice or other source of cells.
3. 40 μm Cell strainers.
4. 1.5 ml Eppendorf tubes, and 15 and 50 ml Falcon tubes, or equivalents.
5. Tissue culture cabinet.
6. FACS machine.
7. Ice.
8. 55 mm Diameter plastic Petri dish.
9. 96-Well polypropylene plates.
10. Heated shaker (kept at 37 °C).
11. Centrifuges.
12. 0.4 % Trypan blue solution (e.g., Sigma Aldrich, UK) and a hemocytometer.
13. Fluorescent FACS-compatible viability dye (e.g., Viaprobe®, BD Biosciences, UK).

3 Methods

3.1 Primary Cell Isolation from Murine Secondary Lymphoid Organs

1. Surgically remove the SLO from a euthanized animal (*see* **Note 1**) and place it in 5 ml of complete RPMI medium that has been chilled previously on ice (*see* **Notes 2** and **3**). Leave the tissue on ice until you are ready to process it (*see* **Note 4**). To ensure data reproducibility, we recommend that SLO from at least three WT animals (and three chemokine receptor KO animals, if available) are prepared and used in all experiments.
2. Transfer the tissues into a tissue culture cabinet and ensure that the following steps are performed in a sterile environment.
3. Using autoclaved sterile tweezers, remove the SLO from the complete RPMI medium and place it into a tissue culture-grade 55 mm diameter sterile Petri dish. Cover the SLO with

2.5 ml of collagenase digestion medium (*see* **Notes 5** and **6**), and cut it into pieces, roughly 2–3 mm^3, using a sterile scalpel or a similar tool. For very small SLO, such as the popliteal lymph node, it is sufficient to disrupt the outer capsule of the node with a single cut. Transfer the diced SLO and collagenase digestion medium to a 15 ml centrifuge tube.

4. Place centrifuge tube in a heated shaker and incubate at 37 °C for 1 h at 750 revolutions per minute (rpm) (*see* **Note 7**).
5. Pass the digested SLO through a 40 μm cell strainer into a 50 ml centrifuge tube. Undigested fragments of SLO caught in the strainer can be gently disrupted with the sterile plunger of a 5 ml syringe and flushed through with a small volume of complete RPMI medium.
6. Add complete RPMI medium to the centrifuge tube up to a final volume of 50 ml and spin at $400 \times g$ for 5 min at 4 °C. Repeat this wash step a second time and finally resuspend the pellet in 5 ml of complete RPMI medium.
7. Determine the number of viable cells isolated by mixing a 10 μl aliquot of cell suspension with an equal volume of Trypan blue. Apply the mixture to a hemocytometer and, according to the manufacturer's instructions, determine the total live cell count (dead cells will appear blue). Resuspend the cell suspension to a density of 1×10^7 live cells per milliliter.

3.2 Fluorescent Chemokine Internalization by Primary Cells

1. Prepare an appropriate volume of fluorescent chemokine (so that you have 50 μl per well of cells) diluted to a final concentration of 25 nM in binding buffer (for biotinylated chemokines *see* **Note 8**).
2. Add 10^6 WT or KO cells (*see* **Note 9**) to each well of a 96-well round-bottomed polypropylene plate (*see* **Notes 10** and **11**). Spin the plate in a centrifuge with a suitable adaptor for spinning plates at $400 \times g$ for 5 min at 4 °C, aspirate the supernatant (*see* **Note 12**), and disturb the cell pellet by vortexing. Prepare additional wells that will serve as competition controls.
3. Resuspend cells in 50 μl of binding buffer containing fluorescent chemokine with or without a tenfold excess of unlabelled chemokine competitor (*see* **Note 13**).
4. Cells are then incubated for 1 h in the dark at 37 °C (*see* **Notes 14** and **15**). Occasionally agitate the cells to keep them in suspension, e.g., 30 min into incubation period.
5. Once the incubation is complete, centrifuge the plate at $400 \times g$ for 5 min at 4 °C, remove supernatant, and disturb cell pellet by vortexing. The cells are now ready for antibody staining and should be kept at 4 °C throughout the remainder of the protocol (*see* **Note 16**).

3.3 Preparation of Fluorescent Chemokine-Labelled Primary Cells for FACS Analysis

1. Resuspend the cell pellet in 50 μl FACS buffer containing 5 μg/ml of Fc block for 15 min at 4 °C (*see* **Note 17**).
2. Centrifuge plate at 400×*g* for 5 min at 4 °C, aspirate supernatant, and disturb cell pellet by vortexing. Cells are then incubated for 15 min at 4 °C with fluorescently tagged antibodies specific for markers of the cell types of interest (*see* **Notes 18** and **19**). Control samples should also be set up in which these antibodies are replaced with similarly tagged isotype control antibodies so that appropriate gates can be drawn during the FACS data analysis to reliably identify specifically labelled cells.
3. Cells are then washed with FACS buffer twice, i.e., re-suspended in 200 μl FACS buffer followed by centrifugation at 400×*g* for 5 min at 4 °C.
4. Upon completion of the wash steps, cells are re-suspended in FACS buffer containing a viability dye (in most cases we use ViaProbe® (10 μl ViaProbe in 90 μl FACS buffer per well)). Incubate the cells for 10 min in the viability dye solution and then add a further 200 μl of FACS buffer (*see* **Note 20**).
5. Data is then acquired for each of the labelled cells using a FACS machine, and analyzed using appropriate software, e.g., FlowJo® (TreeStar, Inc., OR, USA). A detailed description of how to run a FACS machine, set up compensation parameters and acquisition gates, and analyze the data generated is beyond the scope of this protocol. New users should seek assistance from experienced colleagues or the machine manufacturers.

4 Notes

1. We have used leukocyte-rich samples in our studies [5], such as blood or SLO, because they contain large numbers of chemokine receptor-expressing cells and the preparation of single-cell suspensions is relatively straightforward, as described in Subheading 3.1. We would advise using such tissues to optimize the assay for the detection of specific chemokine receptors. However, we have also successfully detected chemokine receptors on leukocytes derived from the lung, gut, skin, and uterus [8] (unpublished results). Any tissue could be analyzed using this assay, provided live cells can be reproducibly isolated.
2. We find that sterile primary cell cultures can be achieved without the need for sterile tissue dissection in a tissue culture cabinet or with sterile instruments provided subsequent culture media contains an antibiotic, such as penicillin or streptomycin or both. However, if antibiotic use is not possible, dissections should be performed in a tissue culture cabinet with sterile instruments wherever possible.

3. Rapidly chilling the tissue on ice prevents excessive cell death by slowing cellular metabolism. We find that cooling tissues in this way and performing each subsequent step on ice or at 4 °C will ensure a high degree of cell viability.
4. In our experience, the SLO can remain on ice for 5–6 h without a major loss of cell viability. However, it is preferable to progress through **steps 1–6** as rapidly as possible.
5. Enzymatic digestion with collagenase is a more gentle process than direct mechanical disruption and results in a better retrieval of larger and therefore more fragile cell types such as macrophages, dendritic cells, and stromal cells. Readers should also be aware that there is a multitude of enzymatic preparations that can be used to disaggregate tissues that may be more suitable than collagenase for the isolation of their preferred cell type(s).
6. If lymphocytes are sufficient for downstream analysis, and larger cells are not of interest, then the reader may choose not to digest the tissue and instead proceed to Subheading 3.1, **step 5**, where the SLO in complete RPMI medium is mechanically dissociated and passed directly through the 40 μm cell strainer using a cell plunger.
7. In our experience enzymatic preparations vary in efficacy from batch to batch; therefore the 1-h incubation period must be taken as a guide. Check the progress of the digestion regularly and avoid where possible extended periods of digestion. Digestion is complete when the medium turns cloudy with released cells, and large pieces of tissue are substantially dissociated. However, this process can be harmful to the cells and remove surface epitopes, which may consequently weaken the antibody staining outlined in Subheading 3.3.
8. Biotinylated chemokines, such as $CCL19^{BIO}$ and $CCL3^{BIO}$, need to be labelled with streptavidin–allophycocyanin (APC) prior to use in the chemokine uptake assay [9–11]. Tetramers of $CCL19^{BIO}$ or $CCL3^{BIO}$ labelled with streptavidin–APC are formed by incubating a cocktail of 6 μg of $CCL19^{BIO}$ or $CCL3^{BIO}$ with 100 μg streptavidin–APC per milliliter of PBS for 45 min at room temperature in the dark. The fluorescent tetramers are then added to the cells to a final concentration of 0.6 ng/μl $CCL19^{BIO}$ or $CCL3^{BIO}$ and 10 ng/μl streptavidin–APC in binding buffer. The basic assay protocol is then resumed at Subheading 3.2, **step 4**.
9. For analysis of rare populations within SLO (<2 % of total), the number of cells stained per well can be increased up to a maximum of 6×10^6 cells without altering any downstream elements of the protocol. When labelling more than 6×10^6 cells, divide samples between several wells with never more than 6×10^6 cells per well. Following completion of the incubation with chemokine, wells can be combined and fluorescently labelled according to the protocol in Subheading 3.3.

10. Due to their highly charged nature, all chemokines, including those that are labelled with fluorophore or biotin, are prone to nonspecific binding to the surface of plastics used in the experiments. In an effort to limit nonspecific loss of chemokines, polypropylene plates are used in the assay. These plates possess superior low adherence/absorbance properties when compared to other alternatives.
11. If desired, 0.5–1.5 ml polypropylene microcentrifuge tubes can be used as an alternative to the individual wells of a 96-well round-bottomed polypropylene plate. All volumes and timings are unaffected by this change; however centrifugation steps, whilst performed at the same speed, duration, and temperature, must now be performed in an appropriate microcentrifuge.
12. In our experience, after a plate has been centrifuged, the supernatants can be safely and easily removed by inverting the plate over a sink. Take the plate in the palm of the hand and in one motion rotate 120° from horizontal and eject the well contents with a flick of the wrist. Return to the upright position and blot the surface of the plate with a paper towel to remove any excess liquid. There should be little or no liquid present in the wells. If this is not the case, then this is an indication that the flick of the wrist was not sufficiently strong. This method is superior to aspiration methods, which run the risk of aspirating away the cell pellets.
13. Competition samples refer to those that contain fluorescent chemokine and unlabelled chemokine competitor. In most cases, the competitor is used to determine which chemokine receptor, if any, is responsible for the uptake seen (refer to Subheading 3.1). A "no uptake" control should also be included, i.e., cells re-suspended in binding buffer alone without fluorescent chemokine.
14. We have found that a 1-h incubation time is appropriate for the detection of every chemokine receptor we have tested thus far (CCR2, 4, 5, 7 and CXCR2, 3) on a variety of leukocytes. Increasing the incubation time will allow more time for fluorescent chemokine to be internalized and will result in increased fluorescence, but nonspecific chemokine uptake may also be enhanced and cell death may increase. If chemokine receptor detection by this method is hindered by high degrees of cellular auto-fluorescence, or alternatively if the cells under study carry low levels of chemokine receptor or chemokine receptors with slow rates of internalization, then increasing the incubation time may help overcome some of these issues. Suitable control samples, incubated for the same period of time, need to be included to verify receptor-mediated uptake.
15. All fluorophores will degrade slowly upon exposure to light; therefore precautions should be taken to limit the exposure of

fluorescently labelled cells to light during, and subsequent to, internalization.

16. Alternatively, labelled cell suspensions can be spun onto glass slides using a cytospin centrifuge at 200 × g for 6 min and analyzed by fluorescence microscopy.
17. Fc block acts to inhibit nonspecific binding of subsequent fluorescent antibodies via their Fc portion, which would otherwise lead to false positives in subsequent analysis.
18. Biotinylated antibodies can also be used to label cells, but only if a biotinylated chemokine is not used. Following incubation for 15 min at 4 °C with biotinylated antibodies, cells are washed twice (Subheading 3.3, **step 3**). Cells are then incubated with fluorescently labelled streptavidin for 15 min at 4 °C and then washed twice, as before. The protocol can then be resumed at Subheading 3.3, **step 4**.
19. Remember not to use APC/AF647 because this channel is already occupied by the fluorescent chemokine.
20. Exclusion of dead cells using a viability dye is required because chemokines stick to dead cells and may therefore give a false positive for chemokine binding.

References

1. Rot A, von Andrian UH (2004) Chemokines in innate and adaptive host defense: basic chemokinese grammar for immune cells. Annu Rev Immunol 22:891–928
2. Hansell CA, Simpson CV, Nibbs RJ (2006) Chemokine sequestration by atypical chemokine receptors. Biochem Soc Trans 34:1009–1013
3. Hansell CA, Hurson CE, Nibbs RJ (2011) DARC and D6: silent partners in chemokine regulation? Immunol Cell Biol 89:197–206
4. Mantovani A, Bonecchi R, Locati M (2006) Tuning inflammation and immunity by chemokine sequestration: decoys and more. Nat Rev Immunol 6:907–918
5. Hansell CA, Schiering C, Kinstrie R, Ford L, Bordon Y, McInnes IB, Goodyear CS, Nibbs RJ (2011) Universal expression and dual function of the atypical chemokine receptor D6 on innate-like B cells in mice. Blood 117:5413–5424
6. Madigan J, Freeman DJ, Menzies F, Forrow S, Nelson SM, Young A, Sharkey A, Moffett A, Graham GJ, Greer IA et al (2010) Chemokine scavenger D6 is expressed by trophoblasts and aids the survival of mouse embryos transferred into allogeneic recipients. J Immunol 184: 3202–3212
7. Nibbs RJ, Kriehuber E, Ponath PD, Parent D, Qin S, Campbell JD, Henderson A, Kerjaschki D, Maurer D, Graham GJ et al (2001) The beta-chemokine receptor D6 is expressed by lymphatic endothelium and a subset of vascular tumors. Am J Pathol 158:867–877
8. Bordon Y, Hansell CA, Sester DP, Clarke M, Mowat AM, Nibbs RJ (2009) The atypical chemokine receptor D6 contributes to the development of experimental colitis. J Immunol 182:5032–5040
9. Comerford I, Milasta S, Morrow V, Milligan G, Nibbs R (2006) The chemokine receptor CCX-CKR mediates effective scavenging of CCL19 in vitro. Eur J Immunol 36:1904–1916
10. McCulloch CV, Morrow V, Milasta S, Comerford I, Milligan G, Graham GJ, Isaacs NW, Nibbs RJ (2008) Multiple roles for the C-terminal tail of the chemokine scavenger D6. J Biol Chem 283:7972–7982
11. Weber M, Blair E, Simpson CV, O'Hara M, Blackburn PE, Rot A, Graham GJ, Nibbs RJ (2004) The chemokine receptor D6 constitutively traffics to and from the cell surface to internalize and degrade chemokines. Mol Biol Cell 15:2492–2508

Chapter 14

Live Cell Imaging of Chemotactic Dendritic Cell Migration in Explanted Mouse Ear Preparations

Michele Weber and Michael Sixt

Abstract

Leukocyte migration through the interstitial space is crucial for the maintenance of tolerance and immunity. The main cues for leukocyte trafficking are chemokines thought to directionally guide these cells towards their targets. However, model systems that facilitate quantification of chemokine-guided leukocyte migration in vivo are uncommon. Here we describe an ex vivo crawl-in assay using explanted mouse ears that allows the visualization of chemokine-dependent dendritic cell (DC) motility in the dermal interstitium in real time. We present methods for the preparation of mouse ear sheets and their use in multidimensional confocal imaging experiments to monitor and analyze the directional migration of fluorescently labelled DCs through the dermis and into afferent lymphatic vessels. The assay provides a more physiological approach to study leukocyte migration than in vitro three-dimensional (3D) or 2-dimensional (2D) migration assays such as collagen gels and transwell assays.

Key words Cell migration, Chemotaxis, Dendritic cells, Dermis, Lymphatic vessels, Interstitium, Skin, Imaging

1 Introduction

During their continuous recirculation between the bloodstream, peripheral tissues, and lymphatics, leukocytes encounter different types of barriers and tissues that they have to negotiate [1]. In the interstitium of the skin, the body's largest organ and its first protective barrier to external pathogens, leukocytes have to migrate through 3D arrays of fibrillar collagen bundles and penetrate the 2D basement membrane (BM) surrounding lymphatic vessels (LVs).

A crucial step in both immune cell homeostasis and activation is the trafficking of activated DCs to regional lymph nodes via afferent lymphatics. Many functional studies have focussed on the migration of DCs in the skin to the cutaneous lymph node [2] but these represent mostly end-point analyses and little information is

Astrid E. Cardona and Eroboghene E. Ubogu (eds.), *Chemokines: Methods and Protocols*, Methods in Molecular Biology, vol. 1013, DOI 10.1007/978-1-62703-426-5_14, © Springer Science+Business Media New York 2013

available on the dynamics and details of DC navigation through the skin's interstitium.

The ex vivo crawl-in assay with explanted mouse ear sheets described here allows the observation of DCs entering LVs in real time with simultaneous visualization of the cellular and extracellular tissue structures. Using this method, it has been shown that DCs require neither integrins [3] nor pericellular proteolysis [4] to reach and enter lymphatics, but rather navigate there through coordinated shape changes and cytoskeletal flow [5].

We provide a step-by-step guide for (1) the separation and immunostaining of ear halves, (2) the generation and labelling of bone marrow-derived mature dendritic cells (BMDCs), and (3) experimental setup considerations for the crawl-in assay and the type of analyses that can be performed.

For the readout of this assay, a standard confocal microscopy setup is sufficient. The crawl-in assay can be employed for studies with pharmacological inhibitors and extended to incorporate two differentially labelled cell populations.

As such, the explanted mouse ear crawl-in assay represents a relatively simple and inexpensive method for the live imaging of chemokine-dependent DC migration in a near-physiological environment.

2 Materials

2.1 Cell Culture

1. Femur and tibia from 6–12-week-old mouse.
2. One pair of forceps (curved or straight).
3. One pair of scissors (small, straight).
4. 1.5 mL Microfuge tubes.
5. Approximate 8 mm ring cut from silicone tubing (8 mm outer diameter, 5 mm inner diameter, 1.5 mm wall thickness).
6. 1 mL Pipette tips.
7. 1× Phosphate-buffered saline (PBS) without $MgCl_2$ or $CaCl_2$.
8. R10 medium: RPMI-1640 supplemented with 2 mM l-glutamine, 100 U/mL penicillin, 100 μg/mL streptomycin, and 10 % fetal calf serum (FCS) that has been heat-inactivated for 30 min in a 56 °C water bath and sterile-filtered through a 0.22 μm pore filter.
9. Petri dishes (sterile, 100 × 15 mm, Greiner 10567571) (*see* **Note 1**).
10. Tissue culture dishes (sterile, 60 × 15 mm, e.g., BD Falcon Ref 353004).
11. Polypropylene conical centrifuge tube (50 mL).

12. Granulocyte macrophage-colony-stimulating factor (GM-CSF): Prepare a 50 μg/mL solution of recombinant murine GM-CSF (e.g., Peprotech, Rocky Hill, NJ, USA) in sterile water and freeze aliquots at −80 °C.
13. Lipopolysaccharide (LPS): Prepare a 1 mg/mL solution of LPS from *Escherichia coli* 0127:B8 (suitable for cell culture, γ-irradiated, e.g., from Sigma, St. Louis, MO) in sterile PBS and freeze aliquots at −20 °C.

2.2 Preparation of Mouse Ear Sheets

1. 6–8-Week-old mice (e.g., C57BL/6 strain).
2. One pair of forceps (straight).
3. One pair of fine forceps (e.g., Dumont #5, straight).
4. One pair of scissors (small, straight).
5. R10 medium (*see* Subheading 2.1, **item 8**).
6. 48-Well tissue culture plate.

2.3 Immunofluorescence of Ear Sheets

1. Primary antibody: Mouse LYVE-1 MAb (Clone 223322) and Rat IgG2A (R&D Systems, Minneapolis, MN) are recommended based on our experience. Prepare a 0.5 mg/mL stock in sterile PBS and store aliquots at −80 °C.
2. Secondary antibody: F(ab $)_2$ anti-rat IgG-FITC (e.g., Jackson ImmunoResearch Europe, Suffolk, UK). Prepare a 0.75 mg/mL stock in water supplemented with 50 % glycerol and store aliquots at −20 °C.

2.4 Fluorescence Labelling of Dendritic Cells

1. TAMRA: 5-(and-6)-carboxytetramethylrhodamine, succinimidyl ester, mixed isomers (C-1171, Life Technologies Ltd, Paisley, UK, is recommended). Prepare a 10 mM stock in dimethyl sulfoxide (DMSO) and store aliquots at −20 °C.
2. Polypropylene conical centrifuge tube (15 mL).
3. R10 medium (*see* Subheading 2.1, **item 8**) without phenol red supplemented with 10 mM HEPES.

2.5 Imaging Chambers

1. Washers (6.4 × 18 × 1.6 mm, stainless steel or plastic).
2. Paraffin: Paraffin wax (melting point >60 °C) or paraffin pellets.
3. Laboratory hot plate.
4. Tissue culture dishes (60 × 15 mm).

2.6 Confocal Microscopy

For 3D time-lapse dual-fluorescence movies, any standard laser scanning or spinning disk confocal microscopy setup equipped with a temperature-controlled heating system, a moveable stage, a 20× objective, as well as 488 and 561 laser lines will be suitable.

Our experimental setup utilized an upright Leica SP5 laser scanning confocal microscope equipped with a custom-built climate chamber (without CO_2 supply).

2.7 Image Analysis

1. Fiji free image processing and analysis software (http://fiji.sc/wiki/index.php/Fiji) with Manual Tracking Plugin incorporated.
2. Manual Tracking Plugin Documentation:

   http://fiji.sc/wiki/index.php/Manual_Tracking
3. Chemotaxis Tool Image J Plugin (Version 1.01) and Notes:

   http://www.ibidi.com/applications/ap_chemo.html

3 Methods

The crawl-in assay requires three preparation steps in two stages: (1) culture of mouse BMDCs over the course of 1 week and (2) preparation and immunostaining of mouse ear sheets as well as fluorescence labelling of mature DCs on the day of the experiment (approx. 2.5 h in total). Each live imaging experiment (i.e., one ear sheet) will take 2.5 h.

3.1 Generation of Mouse Bone Marrow-Derived DCs

1. On day 0, sacrifice one mouse according to local regulations. Carefully release legs from the pelvic bone and separate femur and tibia at the knee joint using scissors. Remove as much muscle tissue as possible from bones by rubbing with paper wipes. Place muscle-free bones into a 50 mL conical tube with PBS. Further processing should be performed in a sterile tissue culture hood.
2. Immerse the bones in a Petri dish with 70 % ethanol for 30 s. Dry the bones in a new dish. Pick up the bones with sterile forceps (thoroughly cleaning with 70 % ethanol beforehand is also sufficient) and cut off both ends with sterile scissors.
3. Place the ring cut from silicone tubing into a 1.5 mL microfuge tube so that it sits halfway inside the tube. Insert a 1 mL pipette tip into the ring and cut it at the top end to a suitable size so that you can still close the microfuge tube. Add 250 μL of 1 × PBS to the microfuge tube. Place femur and tibia of one leg vertically into the opening of the pipette tip. Repeat for the bones of the other leg. Close the microfuge tubes. Centrifuge in a tabletop centrifuge at 1,900 × g for 4 min to harvest the bone marrow. All centrifugation steps can be carried out between 4 and 20 °C.
4. Discard the pipette tip with the now empty bones. The silicone ring can be cleaned with ethanol and reused. Remove PBS and resuspend the red pellet in 500 μL R10 medium by pipetting

up and down. Pool the cell suspensions from both microfuge tubes and resuspend in 10 mL R10. Count the cells and adjust the cell concentration to 2.5×10^6/mL.

5. To each Petri dish (100 × 15 mm), add 1 mL of this bone marrow cell suspension to 9 mL of R10 medium supplemented with 20 ng/mL recombinant GM-CSF. Incubate the cells at 37 °C, 5 % CO_2, 95 % O_2 in a humidified incubator.
6. Three days later (day 3), feed the cells by adding 10 mL R10 medium containing 20 ng/mL GM-CSF giving a total of 20 mL culture per dish.
7. On day 6, carefully remove 10 mL of the culture medium and add 10 mL fresh medium supplemented with 20 ng/mL GM-CSF.
8. Immature DCs are ready to be stimulated on days 8–9 of the culture (*see* **Note 2**). Collect the supernatant containing adherent and loosely adherent cells from the Petri dish. Do not harvest adherent cells. Centrifuge the cells for 5 min at 300 × *g*, discard the supernatant, and resuspend cells from one 100 mm Petri dish in 10 mL R10 + 20 ng/mL GM-CSF. Distribute the cell suspension to two 60 mm tissue culture dishes. The adhesive surface of the tissue culture dish will serve to separate nonadherent mature DCs from adherent immature DCs and some remaining macrophages.
9. To induce maturation of DCs, add LPS to the culture medium to obtain a final concentration of 200 ng/mL and incubate at 37 °C, 5 % CO_2, 95 % O_2 for 24–36 h (*see* **Note 3**). Mature DCs can be distinguished by their characteristic morphology (*see* **Note 4**). Do not use DCs after more than 36 h of stimulation, as they will undergo apoptosis.

3.2 Preparation of Mouse Ear Sheets

1. Prepare a 48-well tissue culture dish with primary antibody diluted in 250 μL R10 medium for immunofluorescence staining of the ear sheets (one well per ear). To stain LVs, add primary antibody against LYVE-1, a marker of lymphatic endothelial cells, at a final concentration of 1–2 μg/mL to the medium in each well (*see* **Note 5**).
2. Sacrifice 6–8-week-old C57BL/6 mice and cut off the ears with scissors. Avoid cutting off too much fur (*see* **Note 6**).
3. Separate ear sheets into dorsal and ventral sides using forceps. To achieve this, place the ear ventral side up and hold it down with one pair of forceps. With the other pair of forceps (fine), grab into the ear and very carefully and gradually pull off the fine and almost transparent uppermost layer of skin. This layer is composed of epidermis and dermis (*see* Fig. 1a–d, Supplementary Video 1, and **Note 7**).
4. Place this cartilage-free ventral ear half dermis side down atop R10 medium containing primary antibody in the 48-well tissue

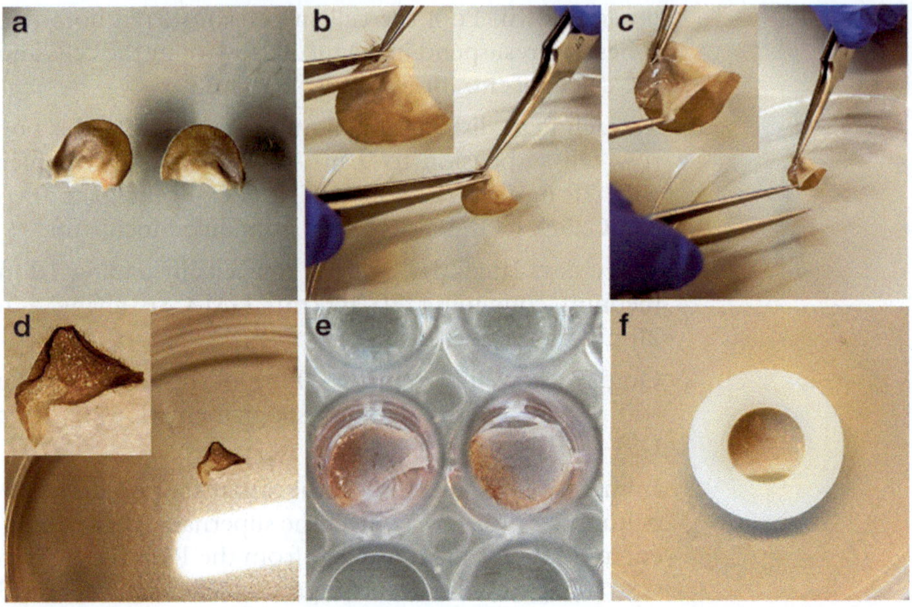

Fig. 1 Preparation of mouse ear sheets for live cell imaging. (**a**) Mouse ears cut off at the base, ventral side up. (**b**, **c**) Mouse ears are separated into dorsal and ventral halves by carefully pulling the two sides apart using dissecting forceps. (**d**) The cartilage-free ventral half is used in experiments. (**e**) Immunostaining of tissue architecture is achieved by floating ventral ear sheet dermis side down on top of R10 containing antibody in 48-well plate. (**f**) Ear sheets are mounted flat, dermis side up in the bottom of a 60 × 15 mm dish, held in place by a washer (sealed to the dish with paraffin—not shown here)

culture plate. The ear sheets should be nicely spread out (*see* Fig. 1e).

5. Incubate in primary antibody for 45 min at room temperature (RT), then discard the medium, and wash three times with 500 μL PBS, gently pipetting up and down each time. Be careful that the ear halves do not flip over.
6. Incubate with secondary antibody (FITC-anti-rat $F(ab')_2$) diluted to a final concentration of 1.5 μg/mL in R10 for 45 min at RT. Cover the plate with aluminum foil.
7. Discard medium and wash again three times before mounting one ear sheet at a time for imaging. Remaining ear sheets should be kept in the 48-well plate with PBS at 4 °C until needed.

Video 1: Preparation of ear sheet (28 s)

Video showing a close-up of the manual separation of dorsal and ventral ear halves using forceps. The ventral side of the ear is facing up at the start of the movie. Available online at http://ist.ac.at/de/research-groups-pages/sixt-group/resources/

3.3 Fluorescence Labelling of Mature Dendritic Cells

1. While ear sheets incubate in secondary antibody, harvest LPS-matured DCs by collecting non-adherent and loosely adherent cells from two 60 mm tissue culture dishes in a 15 mL conical

tube. Centrifuge at 300 × *g* for 5 min. Resuspend the cell pellet in 10 mL PBS and count cells.

2. Prepare the fluorescent labelling solution by diluting an aliquot of the 10 mM TAMRA stock 1:1,000 in 1× PBS (*see* **Note 8**).
3. Pellet cells again by centrifugation. This step will remove most of the FCS present in the culture medium, which can interfere with the cell labelling process. Discard the supernatant and add 100 µL of the diluted TAMRA labelling solution per 10^6 cells. Make sure to resuspend the cell pellet extensively. Incubate the tube wrapped in aluminum foil at RT for 15 min.
4. Fill up the tube with R10 medium and centrifuge at 300 × *g* for 5 min. Resuspend the cells at a concentration of 1×10^6 cells/mL in R10 without phenol red supplemented with 10 mM HEPES.

3.4 Imaging of Cell Migration by Confocal Microscopy

1. Switch on the heating system of your microscope well in advance to reach a stable temperature of 37 °C. Turn on the microscope and switch on the required laser lines. Place a 15 mL tube with phenol red-free R10 supplemented with 10 mM HEPES in a 37 °C water bath.
2. Heat paraffin wax or pellets to 60–80 °C in a small beaker (100 mL size) on a hot plate until paraffin is completely liquefied.
3. Remove ear sheet from the 48-well plate and place it dermis side up in the middle of a 60 × 15 mm tissue culture dish. If the ear borders curl up, carefully spread out the ear sheet using forceps. Place a 6.4 × 18 × 1.6 mm washer on top of the ear sheet, taking care that the ear sheet remains as flat as possible (*see* Fig. 1f and **Note 9**). Seal the washer to the dish by adding melted paraffin around it with a pipette while pressing down the washer with forceps. Allow the paraffin to solidify.
4. Add 250 µL of the 1×10^6 cells/mL labelled DC suspension to the ear sheet. Incubate for 5–10 min in a 37 °C incubator to allow the cells to enter the dermis. Then remove the cell suspension, add 250 µL of preheated phenol red-free R10 medium supplemented with 10 mM HEPES to the ear sheet, and immediately start your imaging experiment (*see* **Note 10**). If more volume is required, a second washer can be placed on top and more medium can be added.
5. To obtain a good overall view of the ear sheet morphology and cell migratory behavior, 20 or 40× objectives are most suitable while 63 or 100× objectives allow higher resolution-focussed views. Choose a suitable location in the mouse ear sheet for imaging: This should contain the initial LVs (LYVE-1+, large

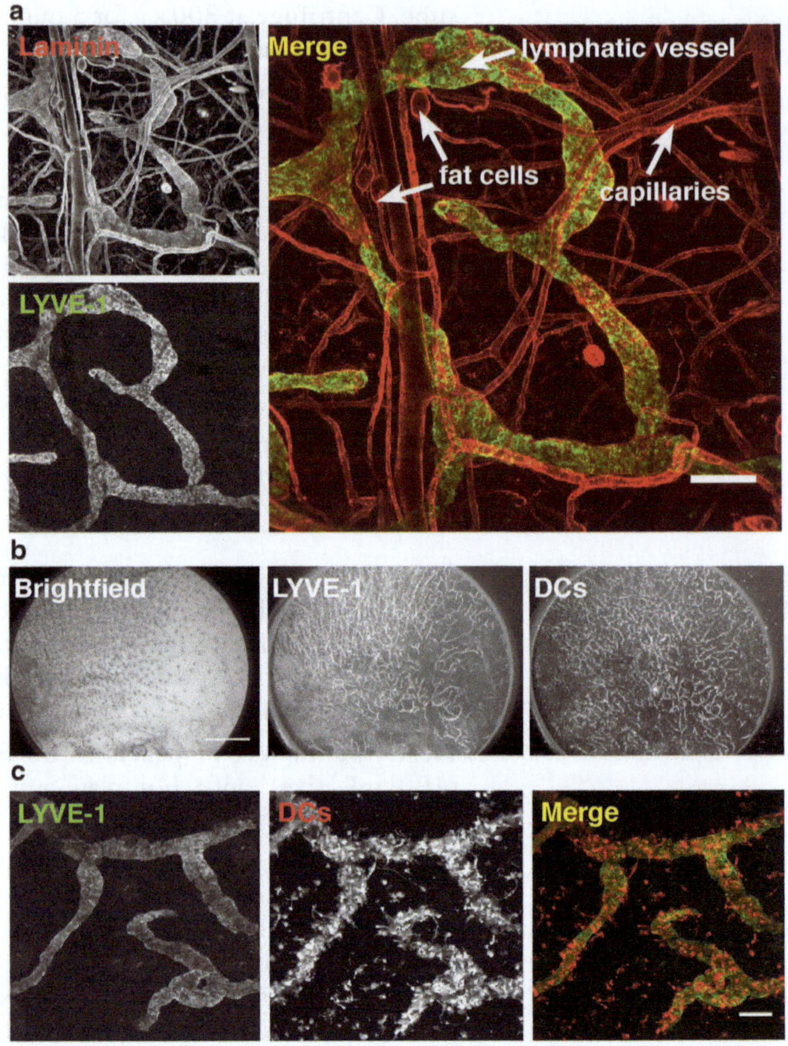

Fig. 2 Anticipated results. (**a**) Immunostaining of ear sheets. Maximum intensity Z-stack projections of confocal fluorescence image series showing immunostaining for BM (Laminin) and lymphatic endothelial cells (LECs) (LYVE-1+), revealing blood capillaries, fat cells, and LVs as indicated. Scale bar: 100 μm. (**b**) Endpoint of ear crawl-in assay. Stereomicroscope images of the entire mouse ear stained for LYVE-1 and incubated with TAMRA-labelled DCs for 120 min. *Left*: Brightfield image revealing hair and hair follicles. *Middle*: LYVE-1 staining of LVs. *Right*: DCs crawled into lymphatics. Scale bar: 2 mm. (**c**) Maximum intensity projections of confocal Z-stack images showing the infiltration of labelled DCs into LVs after 120-min incubation. Scale bar: 100 μm

and irregular diameter, often with blind endings—*see* Fig. 2a and **Note 11**). Acquire Z-stacks encompassing 40–80 μm over a defined time period (up to 120 min). Optical slices of 5 μm and a time resolution of 2–3 min per stack are sufficient for large-scale time-lapse analysis using a 20× 0.5 NA objective.

6. Cells can be seen approaching LVs in a directional manner as soon as 10–15 min after initiating the assay. By 90–120 min, at least 70 % of cells should co-localize with LVs (*see* Fig. 2b, c and **Note 12**).

3.5 Image Processing and Analysis

Image processing and analysis requirements may vary with the ear crawl-in assay depending on the experimental aims. The open-source software Fiji (*see* **Note 13**) is freely available to download online together with relevant supporting documentation. It recognizes a broad selection of file formats and is thus suitable for the range of expected image processing needs and analyses.

1. Confocal microscopy Z-stack series acquired with Leica or Zeiss software can be directly imported into Fiji by drag and drop. Maximum intensity Z-stack projections provide a useful overview of individual snapshots or a complete time series (*see* Fig. 2).
2. Manual tracking of individual cells can be achieved with the Manual Tracking Plugin in Fiji and is sufficient to determine basic chemotactic parameters such as velocity and directionality when combined with the separately downloaded Chemotaxis Tool Plugin (*see* **Note 14**).

For approximate quantification of the number of cells associated with lymphatics, the "Analyze Particles" option in Fiji can be used with a threshold adjusted to recognize the shape of the majority of cells and a region of interest (ROI) drawn manually around the lymphatic vessel staining which can then be overlaid. The morphometric analysis tool in Metamorph software is also useful for the same purpose.

4 Notes

1. It is important to use Petri dishes required for bacterial cultures rather than tissue culture-treated dishes. The latter have an adhesive surface, which is undesirable in the growth and differentiation of BMDCs.
2. Immature DCs can be frozen in a cryotube at a cell density of 5×10^6/mL in 90 % FCS and 10 % DMSO on days 8–9 of the culture and used in experiments up to 6 months later when stored in liquid nitrogen. To stimulate frozen cells, thaw one vial, wash once with R10 by centrifugation, and then add LPS overnight in the same way as described for cells used from the fresh culture. One should however bear in mind that the overall yield of viable mature cells the next day is lower as a consequence of cryopreservation.

3. Alternatively, DCs can be incubated at days 8–9 with other maturation stimuli, such as 500 U/mL TNF-α [6].
4. Mature migratory DCs have the following characteristic morphology: (1) cells lose adhesion to tissue culture plastic 24–36 h after LPS stimulation, and (2) cells in suspension show prominent cell protrusions ("veils"). This phenomenon is dependent on the batch or the lot of FCS used for the R10 culture medium. We recommend batch testing of different sera and their effects on DC maturation. This can be achieved by flow cytometric assessment of surface activation markers such as MHCII, CD86, CD80, or CD40. The expression of these markers does not provide a 100 % guarantee that the mature DCs will have a high migratory potential.
5. Other structures that can be labelled are BMs of both LVs and blood capillaries using anti-laminin antibody, e.g., L-9393 from Sigma as well as other extracellular matrix components [4].
6. Irritation of the skin can cause alterations to the expected steady-state leukocyte migration characteristics; therefore the use of untagged/unclipped ears is recommended.
7. Care must be taken when separating ear halves, as the thin ear sheets can easily rupture and the architecture can be destroyed. In addition to the uppermost layer of skin, part of the underlying cartilage or associated fat cells or both could be obtained during this procedure. This would render the ear sheets unsuitable for an experiment, since the fat cells will prevent successful staining of LVs and penetration of labelled DCs into the dermis. A shiny and more rigid eat sheet (i.e., does not curl up) indicates that a lot of fat cells have been removed. This can be partially avoided by using young female mice for experiments (that appear to have less fat cells) and by taking care not to insert the forceps too deep when attempting to pull off the ear sheet. Slightly curving the fine forceps at their tips may help with the separation of the ear sheets.
8. We have successfully used the following other cell labelling reagents without noticeable effect on intrinsic cell migration: Cell Trace™ Oregon Green (Life Technologies; 3 μM), Cell Tracker™ Orange CMRA (Life Technologies; 5 μM), Cell Tracker™ Red CMTPX (Life Technologies; 5 μM), and Cell Proliferation Dye eFluor 670 (eBioscience; 1 μM). If other cells than DCs are used in the crawl-in assay, the concentrations of labelling reagent may have to be titrated accordingly. One can also combine two populations of differentially labelled cells (i.e., wild type and knockout). In this case it is advisable to perform duplicate experiments with switched cell labelling reagents.
9. The immobilized ear sheets should be spread out as flat as possible to limit problems during Z-stack acquisition and 3D reconstruction. Broader washers with smaller inner diameter

may help to achieve this, although the field of view is then more limited. In our hands, attaching ear sheets with tissue adhesive or super glue has produced very variable results. We have tried this with 3MM Vetbond® and other adhesives, carefully removing excess adhesive with a cotton bud so that it is almost dry before sticking down the ear sheet. This appears to work moderately well when the glue is fresh, but often some of the glue (which is toxic to the cells) would solubilize into the medium, and impair their migration.

10. It is also possible to acquire images with an inverted microscope. In this case, the paraffin-fixed washer is disassembled after the short incubation time that allowed DCs to enter the dermis. 1 mL of R10-HEPES medium is added to the coverslip of a round glass-bottom dish (for example from Matek Corp.) and the ear sheet is placed dermis side down on top. A special dish cover is created by sticking a cap of a 15 mL tube (or any other suitable spacer element) to the lid of the dish. This is put in place so that sufficient space is created between lid and dish bottom that impairs movement of the ear sheet but does not squeeze it (this inhibits cell migration).
11. Best results are generally obtained in the area of the ear with the least hair follicles. When visualizing DC entry into lymphatics, it is important to focus on the initial LVs, not narrower collector vessels (these stain negative for LYVE-1), as DCs cannot enter these vessels under homeostatic conditions.
12. The following variations of the described assay are possible. Rather than performing live imaging, the assay can be employed for simple end-point analysis, for example to determine the effect of an inhibitor or a gene knockdown/knockout on the ability of DCs to reach and enter LVs. In this case, labelled cells are allowed to crawl into unstained ear sheets for up to 120 min after which the ear sheets are washed and fixed (1 % paraformaldehyde overnight at 4 °C or 20 min at RT) and then processed for whole-mount immunohistochemistry [4]. The so-called crawl-out assay represents another modification where the emigration of endogenous $MHCII^+$ DCs from separated ear halves can be assessed under various conditions. To this end, 24-well tissue culture plates are prepared with 0.5 mL R10 medium per well and the cartilage-free ventral ear half placed atop the medium dermis side down. During the course of the next 48–72 h, the ears sheets are incubated at 37 °C, 95 % O_2, 5 % CO_2 and transferred to a new well with medium every 24 h. Emigrated cells can be pooled and kept at 4 °C for end-point flow cytometric analysis and ear sheets can be prepared for immunohistology [4, 7, 8].
13. Fiji is the same as the National Institutes of Health Image J. It comes with a comprehensive package of plug-ins and is continuously updated.

14. Z-stack projections of time-lapse confocal microscopy sequences result in a series of 2D images with cells in different planes of focus. The manual tracking is done in 2D and underestimates cell movement in the *z*-axis. However, the initial LVs of the mouse ear are actually quite flat and span only between 20 and 40 μm in the *z*-axis. Comparisons between manual tracking of cell migration in 2D Z-projections using Fiji and in 3D using more sophisticated methods such as Imaris (Bitplane Inc.) have shown that the 2D tracking method deduces actual velocities and directionalities quite well.

Acknowledgments

We would like to thank Alexander Eichner and Ingrid de Vries for discussion and critical reading of the manuscript, and Mary Frank for assistance with the recording of videos and images in Fig. 1. M.S. is supported through funding from the German Research Foundation (DFG). M.W. acknowledges the Alexander von Humboldt Foundation for funding.

References

1. Nourshargh S, Hordijk PL, Sixt M (2010) Breaching multiple barriers: leukocyte motility through venular walls and the interstitium. Nat Rev Mol Cell Biol 11:366–378
2. Randolph GJ, Angeli V, Swartz MA (2005) Dendritic-cell trafficking to lymph nodes through lymphatic vessels. Nat Rev Immunol 5:617–628
3. Lammermann T, Bader BL, Monkley et al (2008) Rapid leukocyte migration by integrin-independent flowing and squeezing. Nature 453:51–55
4. Pflicke H, Sixt M (2009) Preformed portals facilitate dendritic cell entry into afferent lymphatic vessels. J Exp Med 206:2925–2935
5. Lammermann T, Renkawitz J, Wu X et al (2009) Cdc42-dependent leading edge coordination is essential for interstitial dendritic cell migration. Blood 113: 5703–5710
6. Menges M, Rössner S, Voigtländer C et al (2002) Repetitive injections of dendritic cells matured with tumor necrosis factor alpha induce antigen-specific protection of mice from autoimmunity. J Exp Med 195:15–21
7. Ohl L, Mohaupt M, Czeloth N et al (2004) CCR7 governs skin dendritic cell migration under inflammatory and steady-state conditions. Immunity 21:279–288
8. Price AA, Cumberbatch M, Kimber I et al (1997) Alpha 6 integrins are required for Langerhans cell migration from the epidermis. J Exp Med 186:1725–1735

Chapter 15

Studying Extracellular Signaling Utilizing a Glycoproteomic Approach: Lectin Blot Surveys, a First and Important Step

Jing Cao, Shuzhen Guo, Ken Arai, Eng H. Lo, and MingMing Ning

Abstract

Successful innovative proteomic analysis is highly dependent on molecular biology techniques at the surveying and validation stage. This is because mass spectrometry (MS) analyses of complex samples are limited by their dynamic range for detection—so careful front-end sample preparation, fractionation, and enrichment are crucial to find biologically relevant signals in an extremely complex extracellular environment. Here, we share a very useful and simple front-end surveying methodology—lectin blotting—for proteomic analysis of glycosylation patterns—the most abundant posttranslational modification in extracellular signaling.

Lectin blotting is an effective biochemical technique based on lectin–glycan interactions. It is used to detect and characterize carbohydrate epitopes on protein or lipids. Depending on lectin patterns, specific lectins can be used as tags to enrich glycoproteins for further proteomic analysis. In this method, proteins or lipids are resolved by sodium dodecyl sulfate-polyacrylamide gel electrophoresis (SDS-PAGE) and transferred electrophoretically to adsorbent membranes such as nitrocellulose or polyvinylidene difluoride (PVDF) membrane first. Transferred proteins or lipids are subsequently analyzed by probes including labeled lectins. In this chapter, we provide a detailed methodology of lectin blotting for glycol-proteomic analysis. This method is robust and can be used for complex cell lysates, conditioned media, or human samples to monitor glycosylation pattern following extracellular signal transduction. Here, we will use patient plasma samples post hypothermic therapy—an extremely complex medium—as a tool to describe in detail the technique of lectin blotting for proteomic analysis.

Key words Lectin blot, Glycoprotein, Glycosylation, Glycoproteomics

1 Introduction

Glycosylation is a common and important form of posttranslational modification. It has been reported that more than 50 % of human proteins are likely conjugated with glycans [1]. Glycosylation, especially N-linked glycosylation, is prevalent in proteins destined for extracellular environments, including clinically accessible fluid—plasma. Glycosylation plays vital roles in a variety of biologic processes such as cell–cell communication, cell matrix interaction, adhesion, and protein targeting. Changes of glycosylation status

Astrid E. Cardona and Eroboghene E. Ubogu (eds.), *Chemokines: Methods and Protocols*, Methods in Molecular Biology, vol. 1013,
DOI 10.1007/978-1-62703-426-5_15,

are usually highly related to pathophysiological changes. Thus, the diagnostic and therapeutic potentials in decoding protein glycosylation patterns have been recognized [2, 3] and we feel this process is worthy of investigation in the extracellular milieu.

Recent advances in MS/glycomic technology enables in-depth characterization that utilizes differential lectin binding to broadly characterize complex media such as cell lysates, conditioned media, and human plasma to study in cell–cell signaling. However MS analyses of complex samples are limited by their dynamic range for detection—so careful front-end sample preparation, fractionation, and enrichment are key steps to find biologically relevant signals in an extremely complex extracellular environment. Before in-depth characterization, it is recommended that glycosylation pattern changes with respect to treatment state should be profiled because glycosylation patterns are like "zip codes"—so instead of searching on random streets to pull out just long lists of proteins, we can focus on "neighborhoods" of interest and screen for areas that changed the most due to treatment or disease state. Lectin blotting is an efficient screening method for profiling protein glycosylation status.

Lectins are carbohydrate-binding proteins that may interact with high specificity to a soluble carbohydrate or to a carbohydrate moiety that is a part of a glycoprotein or a glycolipid. Since the first lectin was discovered more than 100 years ago in plants, lectins have been found to occur ubiquitously in nature and have been widely used as an effective biochemical tool [4, 5]. For example, concanavalin A and other commercially available lectins have been commonly used in affinity chromatography for purifying glycoproteins, in lectin arrays and lectin blotting for detecting carbohydrate moieties [6–8]. Here, we discuss a lectin blotting method in detail.

2 Materials

Prepare all solutions using ultrapure water and analytical grade reagents. Prepare and store all reagents at room temperature (unless indicated otherwise). Diligently follow all waste disposal regulations when disposing waste materials. We do not add sodium azide to the reagents.

2.1 Sodium Dodecyl Sulfate-Polyacrylamide Gel Electrophoresis Components

1. SDS lysis buffer (2×): Premix Laemmli Sample Buffer (Bio-Rad) with an equal volume of protein sample buffer containing 62.5 mM Tris–HCl (pH 6.8), 25 % glycerol, 2 % SDS, and 0.01 % bromophenol blue in ultrapure water.
2. SDS-polyacrylamide gel: Precast gels from Life Technologies. Novex® 4–20 % Tris-Glycine Mini Gel (1.0 mm, 12 wells) are polyacrylamide gels based on traditional Laemmli protein electrophoresis.

3. SDS-PAGE system from Invitrogen.
4. SDS-PAGE running buffer: Final concentration of 25 mM Tris, 192 mM glycine, 0.1 % SDS in ultrapure water.
5. Human plasma samples and astrocyte cell culture-conditioned medium (DMEM for control and oxygen glucose deprivation (OGD)/rescue; glucose-free DMEM medium for OGD) concentrated by ultrafiltration centrifugation with 10 kDa cutoff membrane, according to the manufacturer's instructions.

2.2 Lectin-Blotting Components

1. Protein blotting transfer system: iBlot® 7-Minute Blotting System bought from Life Technologies is used to blot proteins from gels to nitrocellulose or PVDF membranes. The iBlot® 7-Minute Blotting System can blot proteins from polyacrylamide gels in about 7 min without the need for additional buffers or an external power supply using disposable blotting stacks with integrated nitrocellulose or PVDF membranes.
2. Tris-buffered saline (TBS; 10×): Final concentration: 1.5 M NaCl, 0.1 M Tris–HCl, pH 7.4, in ultrapure water.
3. Tris-buffered saline-Tween (TBST): TBS containing 0.05 % Tween-20. To make 1 L TBST, add 100 mL of 10× TBS to a 1 L graduated cylinder and adjust volume to 1 L with deionized water. Add 500 μL of Tween-20 (*see* **Note 1**).
4. Blocking solution: 3 % bovine serum albumin (BSA) in TBST. Recommend solution is freshly prepared prior to use. If stored at 4 °C, use within 1 week.
5. Lectin Kit 1 (Table 1, including eight kinds of biotinylated lectins) and streptavidin–horseradish peroxidase [HRP] conjugate

Table 1
Lectins and their primary sugar specificity

Lectin	Common abbreviation	Primary sugar specificity
Concanavalin A	Con A	Mannose
Dolichos biflorus agglutinin	DBA	*N*-Acetylgalactosamine
Ricinus communis agglutinin 1	RCA 1	Galactose, *N*-Acetylgalactosamine
Soybean agglutinin	SBA	*N*-Acetylgalactosamine
Ulex europaeus agglutinin 1	UEA 1	Fucose
Wheat germ agglutinin	WGA	*N*-Acetylglucosamine
Maackia amurensis lectin I	MAL I	Galactosyl (β-1,4) *N*-acetylglucosamine structures
Maackia amurensis lectin II	MAL II	Sialic acid in an (α-2,3) linkage

(Vector lab), Pierce enhanced chemiluminescence (ECL) Western Blotting Substrate, Thermo Scientific CL-XPosure Film, Kodak, M35A-M X-OMAT Processor for imaging and autoradiograph film/cassette.

3 Methods

All the procedures should be performed at room temperature unless otherwise specified.

1. Dilute the plasma samples quantified by Bradford method to 2 μg/μL with deionized water and add SDS lysis buffer (2×) at 1 μg/μL final concentration (*see* **Note 2**). For cell culture medium, concentrated by ultrafiltration centrifugation with 10 kDa cutoff membrane to 15 μL in 30 min, protein concentration is also determined by Bradford method.
2. Heat samples at 95 °C for 5 min to denature proteins.
3. Centrifuge the heated samples at 12,000 × *g* for 5 min to bring down the condensate (*see* **Note 3**).
4. Put Novex® 4–20 % Tris-Glycine Mini Gel into electrophoresis box (*see* **Note 4**).
5. Load samples into the wells (*see* **Note 5**).
6. Run the gel at a constant voltage of 120 V.
7. Stop running the gel when the bromophenol blue front reaches the bottom of the gel (*see* **Note 6**).
8. Rinse the gel in deionized water to remove traces of SDS-PAGE running buffer.
9. To prepare for the trans-blot, open the package of iBlot® Transfer Stack, PVDF mini.
10. Place the gel on the PVDF membrane in the "sandwich" chamber (*see* **Note 7**).
11. Run the trans-blot at room temperature for 7 min (*see* **Note 8**).
12. After the trans-blot, block the membrane (facing up) with blocking solution for at least 1 h at room temperature or 4 °C overnight, on the shaker (*see* **Notes 9** and **10**).
13. Incubate with a 3 μg/mL biotinylated lectin solution in blocking buffer on a shaker for 30 min (*see* **Note 11**).
14. Wash the membrane with TBST on a shaker: twice for 5 min, and then twice for 10 min.
15. Incubate with 0.1 μg/mL streptavidin–HRP conjugate in blocking buffer for 20 min (*see* **Note 12**).
16. Rinse the membrane with TBST on a shaker: twice for 5 min, and then thrice for 10 min.

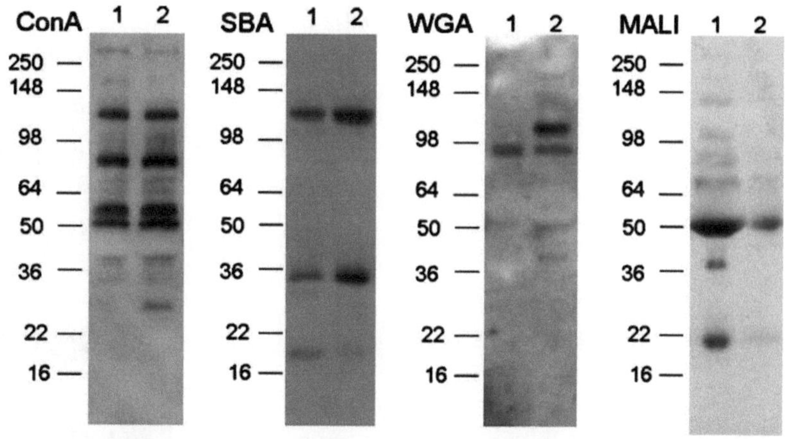

Fig. 1 Examples of lectin blot surveys conducted in plasma. *Lane 1* plasma sample post hypothermic therapy with poor clinical outcome; *Lane 2* plasma samples post hypothermic therapy with good clinical outcome. SDS-PAGE molecular weight standards are marked on the left of each film

17. Prepare one tube for Pierce ECL Western Blotting Substrate solution (3 mL for each gel).
18. Put the PVDF membrane on a smooth, clean surface, such as a glass plate or a parafilm. Pour substrate solution on the membrane and incubate in the dark for about 2 min (*see* **Note 13**).
19. After incubation, absorb any excess liquid.
20. Proceed to the darkroom with the following items: timer, and autoradiograph film/cassette.
21. Put a film on the incubated PVDF membrane to develop. Developing times can vary depending on the relative amounts of glycosylation present.
22. Feed membrane, film, and cassette into a preheated film developer (the machine should be switched on 30 min before film development). Wait for the film to demonstrate distinct signals (Figs. 1 and 2).

4 Notes

1. Cutting the end of a blue 200–1,000 μL pipette tip facilitates aspirating Tween-20 easily because Tween-20 is viscous liquid.
2. SDS precipitates at 4 °C. Therefore, the lysis buffer needs to be warmed prior to use.
3. Centrifuging the samples prior to the run helps remove insoluble debris.
4. Gel gradient percentage is dependent on your requirement.

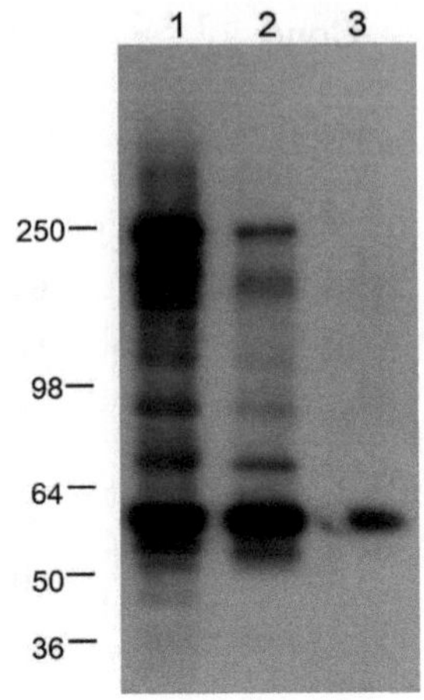

Fig. 2 Example of lectin blot surveys conducted in conditioned cell culture medium and lectin blot with ConA. *Lane 1* control, *Lane 2* post-oxygen glucose deprivation (OGD), *Lane 3* post-OGD/rescue. SDS-PAGE molecular weight standards are marked on the *left*

5. Samples containing multiple proteins require 1–20 μg of protein in 2–10 μL per well. Purified samples containing single or very few proteins require less (0.5–5 μg in 2–10 μL).
6. For certain targeted proteins, it is not necessary to run dye to the bottom. Run SDS-PAGE as appropriate for the protein of interest.
7. Hold the two top corners of the gel with each hand. Lower the bottom part of the gel first on the lower part of the membrane and gently release the gel little by little to lay the complete gel on the membrane. This will prevent trapping of bubbles in between the gel and the membrane. A blotting roller is used to remove any air bubbles between the gel and blotting membrane during the assembly of the stacks and gel.
8. Transfer time can be varied according to the targeted protein's molecular weight. Like in any electrophoresis separation, small proteins (under 30 kDa) migrate more rapidly than large ones and, in the same manner, need less time to transfer from the gel's matrix to the membrane. Seven minutes in this system works well with most proteins; less time is needed to transfer smaller proteins with the iBlot (i.e., 5–6 min). Larger proteins (over 150 kDa) migrate slower than smaller proteins during SDS-PAGE and therefore need more transfer time. The recommended time to transfer large proteins with the iBlot is 8–10 min.
9. In Western blotting, nonfat dry milk is usually used as blocking reagent. However, in lectin blotting, BSA is used as blocking reagent because it is not a glycoprotein. Blocking reagents that contain even small amounts of glycoproteins (e.g., nonfat dry

milk) may bind to the lectins, potentially depleting the available lectin or producing a high background signal. Alternatively, gelatin can be used as a blocking reagent [9].

10. The used gel should be disposed of as hazardous waste.
11. The concentration of biotinylated lectin and incubation time can be regulated according to the type of lectin. Depending on the lectin's carbohydrate specificity, it may be necessary to optimize the working dilution.
12. The concentration of streptavidin–HRP conjugate and incubation time should be optimized according to the type of lectin.
13. Substrate solution should be made fresh. A membrane for mini gel needs 2 mL substrate solution.

Acknowledgement

This work was supported by National Institutes of Health/ National Institute of Neurological Disorders and Stroke grants R01 NS067139 (to MMN) and P01 NS55104 (to EHL).

References

1. Apweiler R, Hermjakob H, Sharon N (1999) On the frequency of protein glycosylation, as deduced from analysis of the SWISS-PROT database. Biochim Biophys Acta 1473:4–8
2. Dube DH, Bertozzi CR (2005) Glycans in cancer and inflammation potential for therapeutics and diagnostics. Nat Rev Drug Discov 4:477–488
3. Cao J, Shen CP, Wang H, Shen HL, Chen YH, Nie AY et al (2009) Identification of N-glycosylation sites on secreted proteins of human hepatocellular carcinoma cells with a complementary proteomics approach. J Proteome Res 8:662–672
4. Sharon N, Lis H (2004) History of lectins: from hemagglutinins to biological recognition molecules. Glycobiology 14:53R–63R
5. Sharon N (2008) Lectins: past, present and future. Biochem Soc Trans 36:1457–1460
6. Yang Z, Hancock WS (2005) Monitoring glycosylation pattern changes of glycoproteins using multi-lectin affinity chromatography. J Chromatogr A 1070:57–64
7. Balcan E, Gümüş A, şahin M (2008) The glycosylation status of murin postnatal thymus: a study by histochemistry and lectin blotting. J Mol Histol 39:417–426
8. Uchiyama N, Kuno A, Koseki-Kuno S, Ebe Y, Horio K, Yamada M, Hirabayashi J (2006) Development of a lectin microarray based on an evanescent-field fluorescence principle. Methods Enzymol 415:341–351
9. Freeze HH (2001) Lectin analysis of proteins blotted onto filters. Curr Protoc Mol Biol Chapter 17, Issue: 1993, Pages: Unit 17.7

Index

A

B

C

Astrid E. Cardona and Eroboghene E. Ubogu (eds.), *Chemokines: Methods and Protocols*, Methods in Molecular Biology, vol. 1013, DOI 10.1007/978-1-62703-426-5, © Springer Science+Business Media New York 2013

H

I

L

M

N

O

P

Q

R

S

T

U

V

W

X

Y

Z

MIX
Papier aus verantwortungsvollen Quellen
Paper from responsible sources
FSC® C105338

If you have any concerns about our products,
you can contact us on
ProductSafety@springernature.com

In case Publisher is established outside the EU,
the EU authorized representative is:
Springer Nature Customer Service Center GmbH
Europaplatz 3, 69115 Heidelberg, Germany

Printed by Libri Plureos GmbH
in Hamburg, Germany